Truckers North
Truckers South

By the same author

Juggernaut Drivers

Truckers North Truckers South

LESLIE PURDON

Old Pond Publishing

First published 1999
Second edition 2007

ISBN 978-1-905523-67-2

A catalogue record for this book is available from the British Library

Published by
Old Pond Publishing
Dencora Business Centre
36 White House Road
Ipswich
IP1 5LT
United Kingdom

www.oldpond.com

Cover design by Liz Whatling
Typeset by Galleon Typesetting, Ipswich
Printed and bound in Great Britain by Biddles Ltd, King's Lynn

Contents

Acknowledgements

I would like to thank my brother Phil for drawing and painting the front cover of my book.

A very special thank you to my lovely wife Pauline, for all her encouragement and hard work on this manuscript.

The back cover picture is by my old friend Alan Spillett. Alan always had a flair for painting, so he enrolled with an art college to study. Eventually his passion for lorries led him to become a lorry driver's mate in 1963, and he then went on to drive eight-tonne lorries himself. The company director Mr Atkins asked him to paint a picture of his house, which Alan did, and later on, his paintings were shown in a Rochester Art Gallery. In 1983 Alan retired from road transport and went back to his first love which was painting. His wife Colleen gave him a great deal of encouragement and stood by him all along. Alan is now one of the leading artists painting anything, but mainly lorries and road scenes.

Today's Driver

The modern LGV driver has an almost invariably thankless task to perform against all odds.

One aspect of the job that has improved dramatically, of course, is the comfort of modern trucks. I remember the early 1970s when we were running ERFs, Fodens and Ford D series, seeing and riding in the first Scandinavian trucks: Volvos and Scanias. What a difference! It was like being transported to a different century – all of a sudden we were in complete luxury. Of course, the rest of the Europeans and the British truck manufacturers followed suit. Unfortunately, probably too late for many.

Equally profound for the driver's physical input has been the effect of curtain-sided bodies as well as palletised fork-liftable and containerised loads. A driver would need to be a fair age to recall the last time he loaded and unloaded bricks by hand. Furthermore, in the past, even when they were mechanically loaded, goods would have to be roped and sheeted, often in high winds and driving rain.

In summary, therefore, it could be said that curtain-siders and shipping containers have changed the face of lorry driving even more than the Scandinavian invasion. However, it will be argued that the decrease in physical input from the driver may not be all good. Maybe the exercise is missed. Take a look at various physiques next time you are in a transport café; I shall not be the judge of that! See for yourself.

One thing is for sure: today's drivers as much as ever before need large amounts of enthusiasm, patience and skill to cope well with what is a rapidly changing, but still very difficult and demanding, vocation.

ADRIAN WATT (Commercial Director)
Kenny Transport Ltd
Peterborough, Cambridge

Lorry drivers and lorry driving have changed enormously over the years. With the advent of the 'Just-in-Time' concept, all those involved in the industry, particularly the driver, have had to adapt to a new and far more demanding role.

Combined with the vast increases in traffic and changes in legislation, today's driver, must also be an ambassador not only for his own company but for the customer as well. This includes looking smart, being helpful, polite, and keen to assist with queries or enquiries, while keeping one eye on the clock for the next delivery!

I do miss the old days of respect, laughter, camaraderie and the more tranquil way of life of the long-distance lorry driver.

LEN VALSLER (Managing Director)
LV Transport
North Fleet, Gravesend, Kent

Introduction

Since 1975, there have been more changes to road transport than at any other time, particularly for the hard-working lorry driver. Trucks have become larger, safer and quieter. All vehicles are now fitted with power steering, air clutches, heating for the cabs and mirrors, air brakes, tilt cabs for access to the engine and two large sun visors. They also have bunk beds with made-to-measure curtains all round, and installed above the beds are powerful reading lights. A safety mirror is secured above the passenger door to reflect onto the front wheel for lorry drivers to see cyclists and other road users.

All cabs today are practically airtight. They have carpets fitted throughout, insulation under the engine bonnet, heating on the ceiling and inside the rear of the cabs.

The control panel has four air clocks, oil, battery and fuel gauges, and roughly fifteen warning lights with symbols to indicate faults. There are other switches, too many to mention.

Some vehicles are fitted with a secondary braking system, such as Jacob's brakes. Others have six-wheel units or four-wheel steering.

When the lorries are driven empty, you can lift an axle by pressing a switch inside the cab to save tyre wear. With their radios and CBs trucks have improved so much that they are now built better than top-of-the-range cars, and that's a fact.

Women can drive juggernauts. There is no physical strength needed, just skill in driving and handling them. With air brakes, trailer boys are not needed.

In bygone days, lorries were not allowed to do any more than twenty miles per hour. In a day's work, a driver would do an average of sixteen miles per hour, so distances which we now treat lightly seemed immense. It meant that the transport driver was a different man altogether. For instance, he would beg, borrow and steal to find work when he was away from home. If he broke down, a company could not send their local van with a fitter to repair the truck; the driver had to get his own help, and do the small jobs himself.

His fellow-drivers would call this sort of person 'a transport man', which meant he was the best. He could drive anywhere, rope and sheet any load. He was capable of getting himself out of any trouble if it involved his work. He would also save his employer pounds.

In his day, roads were narrow, full of potholes, and there was no power steering. Then lorries had gaps between the doors, holes in the floorboards. The only heating the driver had was his overcoat, and the noise from those lorries was unbearable. They had only one windscreen wiper, very small mirrors and one rear light. If they had to pull a trailer, two small triangles were fitted on the back, though a good company would fit an extra rear light. There were no indicators in those days either, so drivers had to use hand signals, and when you drove a lorry and trailer, other drivers could not see you properly because the vehicle was so long.

The transport man drove through every village and town on the map to get to his destinations, as there were no bypasses, dual carriageways or motorways. It was a very hard life. There was no way he could dry his clothes when it rained. He spent more time away from home than modern lorry drivers. He worked out his hours and mileage, then booked his bed in advance daily.

This story is about a transport man and his mate, a trailer boy, in the 1940s. As the years roll by, 'Shay', the trailer boy, learns to drive, and after long years on the road, he ends up driving a modern-day vehicle.

In this book, drivers from all over the country are mentioned, and reading it should bring back many good memories to those who are now retired. To the modern driver, it will offer an insight into how things were. We know that whatever the weather – fog, ice, snow or rain – you will be driving those trucks north and south.

Chapter 1

The Transport Man

I WAS at Watling Street café, Markyate, killing time by tidying up my cab, preparing my bed for an overnight stay. As I sat in the driver's seat, I began to think about how lorries had changed. The vehicle I was driving was a Volvo 88, the most up-to-date truck I had ever handled. It was so easy to drive, a far cry from the trucks of my younger days.

It was 1975, and transport for truckers was changing dramatically for the better. Owner-drivers were buying sleeper cabs, and with the creation of all the new motorways, lorries were going faster and covering more miles. But I was getting very disillusioned. The old days for me were far better, a laugh a minute, and I wished I could be back there.

In my mind I travelled back to 1945, when I was seventeen years old. I had been working for a small local firm as a van boy on a one-ton Commer. The driver I was working with was a good man, but he was no transport driver. I had been trying to get a job on heavy haulage as a trailer boy, but everyone said the same, 'When the grass gets greener' – more or less telling me that I was not ready for it. Then, finally, I got a start working for Banks's Transport of Dartford.

I remember walking down a bumpy unmade road to

the Banks yard from which you could see right across the Thames to Thurrock. It seemed strange to me then, being in Kent and looking at Essex.

As I reported to reception, Mr Banks walked out of his office. He looked at me coldly and said, 'What can I do for you, lad?'

'Are there any vacancies?' I asked in a quiet voice.

'There might be, if you are man enough for the job. Why do you want to work on lorries?' He peered at me. 'It's hard work, you're always away from home – and what would your mother say about that?' Then he said brusquely, 'Before we go any further, turn round and let me look at your shoulders. Mmm, looks to me as if you could do with a meat pudding. Anyway, you haven't answered my question yet. Why do you want to work on lorries?'

'Because my family are all drivers, and I want to get into heavy haulage.'

'Right, lad. I do happen to need a trailer boy. Start Monday, 6 am. Let's have your name and address.'

'Bill Hedley, 50 Miles Road, Gravesend,' I replied.

'Don't forget, I'm the governor,' he warned me. 'The driver will tell me whether you stay or not. Bye, lad.' Taking a quick glance around, I thought to myself what a strange man he was.

On my way out, one of Banks's lorries was coming in, so I walked as close to the edge of it as possible. I noticed it was a six-wheel Leyland Hippo with high side boards; it looked enormous and the roar of the engine sounded really powerful. As he passed me the driver gave me a wink. Banks Transport seemed to be a bag of allsorts, but I loved it. Roll on Monday.

Early on Monday morning I caught the 480 bus from

Gravesend to Dartford. It was a bit of a hike to the yard, but I arrived punctually at 6 am. I was followed in by a fellow on a push-bike.

'Good morning, lad,' he greeted me. 'Old man Banks said I would be having company today. I understand from the governor your name is Bill. Well, mine's Reg, and the first thing on the agenda is a cup of tea.' We walked into the garage through a side gate. In the corner was an old sink with a grubby draining board and a so-called table. The cloth was made of parachute material, and there were a couple of boxes to sit on. Reg made the tea. 'This is Deafy's paradise,' he told me.

'Deafy?'

'Yes, he's the foreman fitter. A split rim blew off a wheel and caught him on the side of the face, leaving him as deaf as a post.' We finished our tea. 'Right, Bill. You wash up and I'll fetch the lorry round.'

As it pulled up, I thought to myself, 'that sounds like an AEC.' Walking outside, I saw it was a six-wheel ERF with an AEC 7.7 engine.

'Jump up, Bill,' Reg called to me. I was full of excitement. No sooner had I closed the door than he let the hand-brake off and we were away. The engine sounded great. I had a quick look through the back window: the lorry had a 21-foot platform which seemed very long to me as we sped down the main road from Banks's.

The wooden cab was rocking, there were gaps at the top of the doors and the old girl was rattling, but I was happy to be working on a lorry at last. 'You won't have to do anything today, young Bill,' Reg told me. 'The driver you'll be working with isn't back. His name's Fred Ruddock, and you'll be his trailer boy.'

By now we had reached fifth gear, and I noticed that Reg was doing the maximum speed the law allowed, which in those days was twenty miles per hour. As we went down through Swanscombe cutting, Reg knocked her out of gear, 'the silent six', and the speedometer crept up to thirty, forty, fifty. 'Bloody hell!' I thought, gripping the door handle. We were really motoring, everything was shaking. 'Keep a look-out for coppers, Bill,' Reg winked at me, smiling. As we dropped speed, he touched the accelerator pedal, revved her up, and dropped the gear stick back into fifth. Now we were purring along, keeping within the law.

Coming into the yard at the Truman Brewery, Gravesend, Reg pulled hard on the steering wheel as fast as he could, turning the lorry round to half its length, then whipped her into reverse. As we were going backwards, he let go of the wheel and the lorry span round like a top, then straightened up as we backed onto the loading bay. Reg put the brake on, then stopped the engine, putting his toe under the accelerator pedal and lifting it up. He looked at me, saying, 'We've got to load empty barrels . . . what are you laughing at?'

'That's good driving, Reg.'

'I always show off when I've got a passenger.' And with that he climbed down from the cab. From that moment, I knew that Reg had taken a shine to me. I only hoped that I got on as well with the driver they called Fred Ruddock. I stood and watched the brewery men load empty barrels; they made it look so easy and in no time at all we had loaded, roped and sheeted. We then made our way towards London.

Now we were back on the old trunk road, the A2,

which had only two lanes. One led to London; the other back to Dover. When we reached the top of Swanscombe cutting, Reg told me we'd soon be stopping for breakfast in the Merry Chest. Within a couple of hundred yards we had pulled in.

As we walked up to the counter, Reg turned round to the drivers and said, 'Who wants tea?' They shouted back, 'We all do.' A driver called Charlie Earl called out, 'Any spare barrels on board, Reg?'

Fred Warren, who worked for Arnold's, added, 'If you have, we'll give you a hand.'

'If I had, I would have dropped them off before coming here and made some money. You bloody lot of vultures would have wanted them for nothing.'

'Shut your row up and get those teas in!' somebody called out. Reg bought two breakfasts and eight teas; it cost about two shillings altogether. 'How much do I owe you?' I asked him.

'Forget it, Bill.' While we were eating, the drivers were reeling out the jokes wholesale. Poor old Reg, he was laughing so much he could hardly catch his breath. What a cheerful man, I thought. As other drivers walked into the café and made their way to the counter, they also asked, 'Who wants tea?'

'We all do!'

'Bastards!' they would laugh as they ordered. Truckers never, ever say no.

'Why have you got a mate today, Reg? Job getting too much for you? You old sod, the old woman must be letting you have too much. If so, share it around with us.'

'Bill's out with me because he's going to be Fred Ruddock's trailer boy.'

Charlie looked at me. 'Fred's a good transport man. What he doesn't know isn't worth knowing.'

'Right,' said Reg. 'We had better get going. We've got a day's work to do, not like you bloody lot sitting on your arses all the time.'

As we made our way to the entrance, they shouted out, 'See you later, Reg.' They were a good bunch. We both jumped up on the ERF and were away. Soon we had reached Blackwall tunnel, which was an experience in itself. Reg drove so close to the lorries coming in the opposite direction that I thought he was going to hit one. In the tunnel there are very sharp bends, and it was an unwritten law that you took it in turns to give way to oncoming traffic.

On reaching the other side, Reg said, 'If you drive too long like that, running your tyres along the kerb, you'll damage them.'

'You have to be good to take a lorry through there then, Reg?'

'You know what your capabilities are, just go through. In other words, young Bill, it's shit or bust.'

As we went through Spitalfields market, there were vegetables strewn all over the place. Porters were pulling heavily laden wheelbarrows which forced Reg to brake, making his life very difficult. As far as the porters were concerned, they had the right of the road, and didn't give a toss about motor vehicles. We then turned right into Brick Lane and Truman's Brewery to load full barrels for the Gravesend bottle plant.

Reg really had to work hard to get the lorry onto the loading bay, a very tight manoeuvre. It did not take us long to unload and reload: the old Londoners at the brewery were good workers and knew Reg well. And what a laugh we had with them afterwards. After knocking back two pints, Reg and I made our way home. Travelling along the road, Reg spoke about the boss, whom he referred to as Banksy. 'He's not the best man to work for, but it's near where I live and I'm home every night. No way would the old girl have me working away from home like Fred and the rest of the nut cases.'

Looking around the vehicle, I realised that the varnish on the timbers had started to fade. 'How old is this lorry, Reg?'

'It came home in 1936,' he replied. 'It's nearly ten years old.' The ERF had a vacuum gauge just under the steering wheel, a couple of clocks on the dashboard, and that was it. It didn't take us long to tip at Gravesend and arrive back at the depot, where Reg pulled onto the diesel pump. 'From now on, Bill, this will always be your job when you start working with Fred.'

Reg then proceeded to show me what to do. He turned the handle round, put one gallon in, re-wound the handle, turned it back and put another one in. 'Got it, Bill?' he asked.

'Yes,' I replied.

'And don't forget to check the bloody oil. It's your responsibility from now on.'

Parking the lorry up alongside the others, we made our way back to the garage for a cuppa. Reg introduced me to some of the drivers who were already there: 'This is Dave Evans, the man from the valleys, and the sooner he goes back to Wales the better. This is Bill Warnett –' a friendly hello – 'and this is Sharpy. He's got a bolt missing. He's never been the same since he left Dunkirk.' We all laughed.

Then in came Deafy. 'What's going on?' he asked.

'Calm down. We've just heard that Banksy's getting rid of you,' said Sharpy.

'What's he saying, what's he saying?' Deafy kept repeating. Sharpy went up to him and shouted in his ear, 'We've heard on the grapevine that Banksy's getting rid of you.'

'What!' Deafy shouted, blowing a fuse.

They pointed at me, saying, 'He's the new replacement. You're too old, and not with it. The office girls say that you're nothing but an old bodger costing the company too much money.'

Next thing, Deafy was gone and banging on Banksy's door. The drivers were killing themselves with laughter. They said, 'He's the best fitter in Kent but so easy to wind up. Banksy thinks the world of him.'

Dave looked at me, saying, 'Put the kettle on, Boyo. We'll have some more tea.' Then in walked Banksy, and I cringed with embarrassment.

'Why do you keep winding Deafy up? He threatened to put one on me. He hates my guts, and most of it's through you drivers. One of these days I'm going to sack you, Sharpy.'

'Sack me now, Governor, and I won't have to drive that bloody old clapped-out Vulcan. And while we're at it, why don't you take a long walk off a short pier!'

Banks retorted, 'Your days are numbered.' He walked out, slammed the door behind him and left us all laughing.

Deafy was getting excited again: 'I told him, the old bastard. I told him!'

One of the lads said, 'Let's go and see what our orders are for tomorrow. We're not like blue-eyed Reg here, knows his orders every day – piss-artist of Truman's.' Dave Evans ran his fingers down the list. 'Oh, you're with Fred Ruddock tomorrow.'

No sooner had he mentioned Fred's name than an eight-wheeler Atkinson and trailer pulled into the yard. It looked handsome. Fred Ruddock climbed down from the cab, and they all gathered round to have a chat. His trailer boy, meanwhile, walked straight into the office for his cards. He had had enough of being away from home, and from what the drivers were saying he was not cut out for road transport.

I stood studying Fred, without his noticing. He was about 5 ft 7 in tall, broad-shouldered, an extremely good-looking man of around thirty-seven. He had black hair, very dark brown eyes, and was wearing a leather bomber jacket with a fur collar – very distinguished looking, in my opinion. After talking with the lads, he said, 'Is toss-pot in?'

'Yes, the old bastard's in,' they replied.

As Fred entered the office, his ex-trailer boy was coming out. Without saying a word to anyone, he walked up the lane with his suitcase and was gone.

I walked around the lorry. It was blood red, and the chassis and wheels were dark green. It had a square cab with two large towing hooks, a spring-loaded starting handle with a brass end, and an aluminium radiator with a large letter 'A' encircled by a ring. It also had two big headlamps and side lights on brackets.

As I was half-way down the lorry, I first became aware of the distinctive smell which big trucks give off. It stayed with me for the rest of my life. From that moment I was hooked. I had fallen instantly in love with a truck.

The lorry and trailer was about 45 foot overall. I wondered how on earth Fred managed to drive around the country from John O'Groats to Land's End – it was all single carriageway, and you went through every town on the map. You had to be really special to drive an eight-wheeler and trailer: a true 'transport man'.

I opened the passenger door and peeped in. I spotted the large ratchet hand-brake, a new modification. Some lorries just had hand-brakes like cars, but much larger, and in others you turned a wheel. I could see it was a new lorry.

Driver Ruddock came out of the office, followed by the governor. Fred stopped and turned around, his face like thunder. 'I'll have a week's notice, Governor.'

The old man shouted, 'You're not leaving. I'm sacking you!'

Fred shouted back, 'That's just what I wanted, you old bastard. If you sack me, I can claim dole money.'

The old man changed completely. 'Don't be like that, Fred. You're tired. Come into the office. We'll go through your expenses again.'

Fred stared at him. 'Governor, I've never known anything like this place, it's a mad-house.'

Then Sharpy appeared and said to me, 'I'll see you in the morning, young Bill. You'll be all right with Fred, but don't upset Deafy by saying you're his replacement, like you did before.' Grinning to himself, he walked away.

Fred looked me up and down. 'I take it you're my new trailer boy. If you're like the other young waster you won't last the week out. He thought that being a trailer boy meant being chauffeured about all day. Come on, for a start we'll derv up.'

I walked to the pump and waited while he drove round. I turned that bloody handle round forty-five times then, without his asking me, climbed up into the cab, moved the blankets and lifted up the engine bonnet to check the engine oil. In the chassis was a beautiful 6LW Gardner engine. Putting everything back as it was, I closed the bonnet.

'Right lad, do you know how to work a ratchet hand-brake?' Fred demanded.

'Yes, my dad's an eight-wheel driver, and I've watched him operating one.'

'Show me.'

I stretched forward, grabbed the hand-brake firmly and pulled it towards me a couple of times to take up the slack until it would not go any more. Then I pushed it forward, back into its original position.

'That's it,' Fred said. 'Now let it off.' I brought the hand-brake forward and threw it back, hard. 'Bang!' it went when it was released.

'That's fine. Tomorrow, fetch a suitcase and I'll see you at six in the morning,' said Fred with a smile. 'Turn up, because I can't move without you. That's the law on lorries with trailers.'

Chapter 2

Fred Ruddock Changed my Name and Way of Life

I ARRIVED at 5.30 on Tuesday morning feeling very anxious about Fred Ruddock. Then in came Sharpy. 'Don't stand there, Bill, let's go to the garage and have a cup of tea. You never know our luck, Deafy might come in early and we'll have a giggle.' I could see then that Sharpy loved to wind people up – not to be nasty but purely for a laugh.

'Do you realise, Bill, you'll be away from home for weeks at a time now you're working with Fred, because he's a typical transport man. If you want that way of life yourself, listen well and you'll learn a lot from him. I'm off to Dover now, might see you in ten years' time,' he said, grinning all over his face.

A little while afterwards Fred arrived. 'You're nice and early, Bill. Well done. Keep it up, but not your pecker. We'll get cracking.'

We walked to the lorry with our suitcases. 'Do you know what make of lorry this is, Bill?' Fred asked.

'Yes, it's an eight-wheel Atkinson with a Dyson trailer, Gardner LW engine and a David Brown gearbox.'

'That's bloody good. I can tell you're interested in

lorries. By the way, you didn't tell me what the chassis number was.' I just laughed, thinking that at least he had a sense of humour.

Once inside the cab I put my suitcase on the floor and Fred put his up on the engine bonnet in front of the rear window. He hit the starter button and she flew into life. Fred pulled a log sheet out of his folder. 'What's the date today?'

'7th March,' I replied.

'Thanks.' He then proceeded to ask my full name: William Hedley.

As we travelled down the road to load up at the Imperial paper mill, Gravesend. I felt really comfortable. This lorry was new. The Gardner engine had a sound that was quite unique, coming from a best-pulling diesel engine, the most reliable ever made in Britain. The cab had varnished timbers, and the whole layout was different from the ERF's: the dash panel was under the steering wheel, slightly to the right, and the ratchet hand-brake was more or less lying on the floor. To me it was beautiful.

Soon after we arrived, the driver of the overhead crane had us loaded. We roped, sheeted and were ready to roll.

'We'll make our way to Ginger's in Barnet for breakfast,' said Fred. 'Everyone calls it the Hole in the Wall.'

'Okay, Fred.' Who was I to argue?

Going up Swanscombe cutting, the Gardner was barking loudly, and I asked, 'How much weight have we got on, Fred?'

'Nineteen tons,' he replied, though the noise from the engine made it hard for me to hear him.

As we headed towards London, other lorry drivers flashed their lights: they all knew Fred. Just as we were coming towards the Dover Patrol public house, they started to give him the thumbs down. He looked at me and told me that the 'wooden-tops' were about. As we passed the pub I looked into the car park and realised what he meant.

'Fred! The coppers are in there!'

'Crafty bastards. They've got nothing better to do,' he replied. I glanced across at the speedometer: it registered twenty miles per hour.

As we drove down Blackheath hill, Fred braked hard. The roar from the engine was deafening. My ears were ringing.

'Take up the slack, Bill!'

'Eh?'

'Take up the slack, and give me three notches! Wind that bastard up and be quick about it!'

Fred brought her to a halt. 'Not bad for a learner,' he said.

'Eh?'

'Why do I have to repeat myself half a dozen times?' he shouted. 'Bloody hell, are you deaf?' As the lights changed from red, we started to laugh as we both knocked the ratchets off. The noise from the drums was terrible, they were literally screaming, and you could actually smell the brake linings inside the cab.

As we drove through Deptford the traffic was heavy; even the trams were overtaking us. I thought to myself how funny they looked plodding along on their little iron tracks. Now and again they weaved in and out, sometimes stopping in the middle of the road to let passengers off, and people ran out in front of us at times, which was quite scary. If only they'd known the weight we had on board and the distance it took to stop a lorry and trailer, they would have thought twice before taking a chance, no matter how much of a hurry they were in.

Travelling up the Old Kent Road, I could see there was still a lot of bomb damage from the war. It was a real eye-opener driving through that part of London.

When we arrived at the Hole in the Wall, a few of the drivers from Arthurells were already tucking into a hearty breakfast of sausage, egg, bacon, beans and fried bread. Just the sight of them made us feel ravenous. 'Let's sit over there with them,' said Fred.

When we had ordered, Fred introduced me to Ted Patterson and George White, who were from Medway in Kent. 'Where are you two off to now?' Ted asked.

'Leeds,' replied Fred.

'Oh, we're going to the London Brick company to load. We'd better get going, lads, we're running a bit late. Well, nice seeing you again, Fred. Bye, Bill.'

'Did you enjoy your breakfast?' Fred asked.

'It was bloody handsome,' I replied.

'Good. You can pay for us.'

I was pretty embarrassed, because I wasn't that flush with money. Seeing the look on my face, Fred laughed and walked up to the counter to pay. Then we made our way to the Great North Road on the A1.

It was the first time I'd been north of London, and I didn't realise there were so many different makes of lorries: steam, petrol, diesel, ex-military vehicles converted for civilian use. Most of them were overloaded and precariously roped up with wire, string, etc. Some of these lorries should never even have been on the road, and the drivers had a really hard life. They never knew if they would get to their destinations or not.

Despite being the main highway, the road was very narrow in places, so narrow that you could almost touch the oncoming traffic with your hand. It was a long old haul, but we eventually arrived at Tony's in Grantham with the Gardner engine still ringing in my ears. We parked alongside one of Bob Davidson's lorries from Aberdeen, an eight-wheel Thornycroft.

As we walked across the lorry park, one of Watt Brothers' trucks pulled in. Fred looked at me and said, 'That's old Kennedy. We'll have a good time tonight.'

After we'd showered and shaved, Fred shouted to me, 'Shay, shit in your boots, son. Make yourself taller! Because, as from tonight, you're eighteen.'

'Don't be like that. I'm not that short!' I grumbled.

'Only teasing. But you'll have to get used to having your leg pulled, otherwise you won't last the week.'

After finishing dinner, we all decided to trundle off to the local pub which was frequented by drivers. The landlord and his regulars were great, with a terrific sense of humour.

After downing his first pint, Fred walked over to the piano and started to play. He was good and that really got things moving. 'Shay,' he said, 'when they're all half-pissed, I'll play "There's an Old Mill by the Stream". That's when you take a glass round the bar and say, "Don't forget the pianist." All right?'

'Okay, I'll do that.' Then I hesitated. 'Fred, why do you keep calling me Shay?'

'Because you keep saying "eh" whenever I speak to you. So from now on, Shay's what your nickname will be. Oh, and by the way, when we get home I'm having your ears tested, otherwise we'll have two Deafys in the firm, and that won't do. The first trailer boy I had with me was half-blind, the second had half a brain and you're bloody deaf. Banksy must have employed you lot from the local nut-house.'

The top of the piano was laden with beer. Fred shouted over to a driver called Big Ian: 'Tell the lads to drink up, the sooner it's drunk the quicker the glasses get filled. But watch young Shay for me, he's under age. Can't have him being sick. It wouldn't go down too well with the landlady!' What a good night this was turning out to be – a laugh a minute!

Everyone was singing and having a great time when this vivacious looking woman walked into the bar. She was very curvaceous with beautiful blonde hair, and she seemed to be known to all the drivers as Sandy. Who'd have believed they all knew such a stunner! She soon started talking and mingling with the drivers.

Big Ian was a powerful man who had arms like a gorilla and hands like shovels. Suddenly he picked Sandy up and turned her upside down, showing off her frilly underwear, suspenders and stocking tops. Everybody was in fits of laughter.

Fred had to stop playing because he was laughing so much. Then, out of the blue, Roy, who was from Liverpool, pulled Sandy's knickers down. I couldn't believe my eyes – there in front of me was a two-inch penis! This was all new to me.

The whole pub was in uproar; everyone had known about Sandy except me. If Fred and the others had known what I had been thinking when 'she' walked into the bar, I would never have lived it down.

When Big Ian put Sandy down, she quickly pulled up her knickers and, looking up into his eyes, said in a sexy voice, 'My, you are so big and strong, I like you!'

'Geroff, you dirty bastard,' he replied.

They all teased Sandy mercilessly, but she loved being the centre of attention and reeled off jokes like there was no tomorrow.

She grabbed a stool and crossed her legs, saying, 'What do you think of this one, boys? I went to the fair with my boyfriend Jamie the other day, and had a ride on the big wheel. It had all pretty lights on it, and it was massive – the wheel, I mean. Jamie wouldn't come on the wheel, so I went on my own.

'I went round once, then twice, but as I was going round for the third time it started shaking violently. Then it collapsed to the ground, and in the distance I could hear Jamie's voice shouting, "Sandy! Sandy!" He pulled the debris from my face and said, "Are you hurt, dear?"

'I looked at him in dismay and said, "Of course I'm hurt, I went round three times and you didn't even wave once!"'

Fred told me afterwards that the teasing was all in good fun, never malicious.

Unfortunately time was running out. Fred started to play 'There's an Old Mill by the Stream', which was my cue to grab a glass, take it round and say, 'Don't forget the pianist.'

Fred earned well that night.

We were on our way back to the digs, trying to keep our voices down, when suddenly out of the blue Big Ian picked up a dustbin and threw it down the end of our garden. The noise of the dustbin could have woken the dead. As our digs were opposite a cemetery, I imagined all those bodies sitting up and saying, 'What the hell was that?'

'Run for it,' said Fred. 'He's a nutter.'

Six of us made for the door of our digs at once. What a squeeze; we must have looked like the Crazy Gang. On reaching the top of the stairs we all undressed quickly and jumped into bed, playing the innocents, giggling like little boys.

The next thing I knew, Big Ian was shaking me. 'Time to get up, laddie.' I awoke with a start, still hearing the sound of the dustbin exploding at the bottom of the garden. At breakfast, we talked amongst ourselves. I overheard Kennedy say, 'What an evening we all had last night!'

'One of the best,' Big Ian replied.

When it was time to go, all the drivers thanked Fred for the beer, said their goodbyes and made their way to the lorry park.

While I was checking the engine oil, Fred gave the lorry and trailer the once-over. When we were inside the cab he started her up. We both released the handbrakes, and we were on our way to the Great North Road. After a good three hours we arrived at the other side of Doncaster, stopping at a café for a cup of tea. I stood up to pay, but Fred stopped me. 'I'll do that, Shay. Our beer money will pay for it.'

'But you paid for the breakfast at Barnet,' I protested.

'If and when the money runs out, you can pay. Until then, shut your cackle and sit down.' Then, as always, Fred handed me half the *Daily Mirror*. 'Now, let's see what Jane's doing today. I wonder what she looks like fully dressed? Still tasty, I would imagine.'

‘Fred, you know what?’ I said, ‘I can still hear the LW ringing in my ears.’

‘Oh, after two or three weeks you’ll get used to it. The time to worry is when you go for a piss and it comes out as diesel!’

Walking across to the transport park, I noticed several different makes of vehicles. There were Dennises, Fodens, AECs, Bull-nosed Bedfords and Morris Commercials. Many had petrol rather than diesel engines. No big deal, you might think, but I was smitten. I knew at that moment that trucking was going to be my way of life.

All I wanted to do was climb inside the cab of the Atkinson; that distinctive smell it gave off was getting into my blood, I loved every minute of the job.

‘Let’s go, Shay.’

‘Okay, Fred.’ We thrust those hand-brakes forward with great force. The Gardner engine sounded terrific as we pulled out of the lorry park. Fred put her into fifth gear, and drove off at about twenty miles per hour.

She was a lovely lorry, and when we closed the doors you would have thought it was a train door shutting. They were really well built in those days. The accelerator pedal was in the middle of the floor, and above was a large round reading light situated in the ceiling. There was also a nine by nine inch flap in the roof for ventilation. The cut-out lever to stop the LW was on my side of the bonnet, and if you did not keep hold of it long enough the Gardner would restart. If you stalled while you were in gear, it would fly backwards, making a terrible noise and emitting black smoke. But it was still an engine at its best.

We did not stop again until we reached Leeds. Going through Rothwell, the drivers coming in the opposite direction kept flashing their lights, giving us the thumbs down, 'Wooden-tops are about! Keep your eyes open, Shay.'

'There's one, just further up on your right, sitting on his motorcycle.'

'They've got nothing else to do,' said Fred scornfully.

'You haven't got much time for coppers, have you, Fred?'

'I'm afraid not. But that's just me.'

The speedometer now read eighteen miles per hour. On arriving at Leeds, we reported to Arnold's Print Works, where it did not take us long to off-load. As we were leaving, the manager rushed out of his office shouting, 'Would you do me a favour and deliver a load to Preston?'

'It depends,' Fred answered.

'Go on. I'll make it worth your while,' pleaded the manager. Fred hesitated. 'What are your rates?'

'I'll pay you well, don't worry.'

'All right, let's discuss it.' Fred disappeared into the office.

Ten minutes later Fred and the manager came back out. I heard the manager say, 'Thanks for helping us out, driver,' as he handed Fred half a crown.

Fred told me we had to back the trailer onto the loading bay, and after about an hour we had loaded it up. We then made our way to the town centre. 'That was a stroke of luck, getting a load to Preston,' he grinned, rubbing his hands together and singing 'We're in the Money'. 'Old Banksy'll be pleased with this load – no dead mileage.'

When we got to Whitehall Road, Leeds town centre, carrying our suitcases, we made our way to our digs.

We awoke early the following morning as we wanted to leave early. Walking across the lorry park, we noticed a driver was having difficulty starting. Fred shouted across to him, 'Want a hand, mate?'

'Please, I think my batteries are flat.' Fred put his hands on the starting handle, gave it a quick jerk and the engine burst into life with a roar. The driver thanked us profusely, telling us he was running late. I suddenly realised what a powerful man Fred Ruddock was. Not many drivers could start a vehicle like that.

Travelling along the A646 heading for Preston, we stopped at a transport café in Burnley. I asked Fred how many tons we were carrying, and he told me nineteen. 'I'm fully loaded today,' he said, 'and it's a gruelling drive. Some of the hills are really steep. I'm not too worried about going up. Going downhill and stopping the bloody thing is the hardest, and trying to keep the drums as cool as possible, using the gearbox more and the brakes less to keep the vacuum up.'

As usual Fred bought the *Daily Mirror* and ordered the tea. Before he could mention Jane, I said it first, grinning at him. 'Are you taking the Micky?' he demanded, laughing.

'You could say that,' I grinned.

'Come on, piss-taker, let's get going. The sooner we get tipped the better I shall like it.'

'Does that mean we're going to do over ten miles an hour up those hills?' I asked.

'I'll have a bloody good try, anyway,' said Fred, adding, 'And don't be so sarcastic. It's the lowest form of wit.'

'Getting jumpy now, are we, or is it just stress?' I joked. 'You're coming up to your forties now. It's the funny age.'

'That's it, next café we come to you're buying. Bloody forties, indeed!'

I could see the strain on Fred's face as he tried to manoeuvre the eight-wheeler into the small space at the print works. It was a real headache for him; the trailer just refused to go in. 'Unhook the trailer, Shay.'

'Right oh.'

'I'll drive out and come back in again, then you hook the tow-bar onto the front of the lorry, and I'll try pushing the trailer back as far as I can into the loading bay. If it wasn't for the bend in the entrance I would've done it no problem.'

The other drivers were admiring the way Fred handled the situation. 'Well done,' said the charge-hand, 'you've saved my men a lot of work.' It was true what Charlie Earl had said in the Merry Chest: Fred really was a good transport man.

After unloading, Fred walked off, saying he'd give Sutton's a ring to see if we could get a return load. After five minutes he was back. 'We've had some luck, Shay. They've given us a load out of Blackburn.'

Preston was one of the main stop-overs for drivers. That night we stayed at the Olympia café, whose clientèle consisted mainly of Scottish lads, and after our usual evening meal, we changed into casuals and went to the local, which naturally had a piano. I asked Fred if he'd been there before. 'Many times,' he replied. After our first drink, he went over to the Joanna and started to play.

The Scottish lads loved it. One of the drivers from Dunfermline, Jock, walked over to Fred and started to sing; he was really good. Naturally, they were Scottish songs but Fred knew them all. I realised at that moment that it didn't matter whether you came from England, Scotland or Wales, all the drivers had a similar sense of humour. When a crowd of them were in a pub together, it got rather rowdy at times, but it was all in good fun. They didn't interfere with anybody unless people butted in and tried to make trouble. Then it was – beware!

As the landlord shouted, 'Time gentlemen, please', Fred started to play his usual 'There's an Old Mill by the Stream'. I quickly took a glass round the bar, not forgetting to say my party piece: 'Don't forget the pianist.' When the customers had gone, I tipped the money out onto the table. Even the landlord had thrown in two shillings.

'Fred, I've collected twelve and six.'

'Well done, Shay,' beamed Fred, and proceeded to give me two and six. He gave Jock a dollar and kept one for himself.

On our way back to the digs Fred started singing 'We're in the Money'. I joined in too, feeling very happy. 'I'll book an early morning call for tomorrow, Shay. You know what they say. . . .'

'What's that?'

'The early bird catches the worm. Haven't you heard that one?'

I hadn't, but I could see his point.

Next morning we stopped just outside Preston at a place called Sanlesbury Bottoms. Fred pulled up on a grass verge and jumped down from the cab. 'Don't just sit there like a prize prick. Get the rope, Shay, and follow me.'

In a field to our right were two horses. Fred patted one gently and gradually slipped the chain off his neck, replacing it with a rope. He did the same to the other, making sure the ropes were tied to stakes. Fred suddenly shouted at me to run, and I ran like a bat out of hell, dragging the chains behind me and quickly throwing them up onto the back of the cab. Fred jumped up behind the wheel and drove away as fast as he could. 'I should have taken the bloody horse-shoes as well. I could have made toggles out of them,' cried Fred.

After a while we pulled up alongside a building site. We wanted to see if we could find some pieces of metal. We needed one end thicker than the other so when it went through the links of the chain the metal wouldn't go right through. After looking around for a while I found two pieces. 'What've you got there, Shay? Looks all right to me. We might be able to get some more at the steel-works.'

On leaving the site, I noticed Fred had a four-foot scaffold pole under his arm; I couldn't believe my eyes. 'Bloody stupid, leaving that pole on the ground,' he muttered. 'Somebody could have fallen over it and hurt themselves. Come on, we'd best get out of here. Folks will get the wrong impression and think we're thieving.'

On arriving at Blackburn Steel, Fred didn't waste much time in cadging timbers. He managed to collect about seven – four for the lorry and three for the trailer.

The manager told us to load eighteen tons of steel for delivery to Strood in Kent. The crane driver soon had us loaded. Fred pulled the lorry to one side, then we threw the chains over the load. We put the metal bars through the links of the chain. Fred put the scaffold pole onto the end of the metal, pulling it down. It cracked as it tightened up. Keeping the bar down, Fred tied the metal to the chain so it would not come undone, then removed the pole, next doing the same to the trailer. He certainly knew his job. 'That's pretty nifty, Fred,' I said, impressed.

'When you haven't got toggles, you have to compromise. Oh, by the way, we'll be stopping at Durose tonight near Newcastle-under-Lyme. Durose is a transport café just for lorry drivers, open twenty-four hours a day. It does B & B too. It was lucky, getting a load for Kent, Shay.'

'Yes, it's only six miles from where I live, Fred.'

'I'd never have guessed that in a million years,' he retorted.

'Now who's being sarcastic?' I laughed.

'Just getting my own back. Anyway, enough of that, let's get our fingers out. I'll get parked up for the night because I just happen to know of a nice little pub in Newcastle.'

'Oh no, that means I've got to go round with the glass again!'

He laughed, calling me a little bastard. 'Okay, I'll go round with the glass and you play the piano, Shay.'

'If you want to hear a professional that's entirely up to you,' I said with mock arrogance. 'It's bad enough watching your bad driving all day.'

'I've got a headache coming on, Shay. Just shut up.'

Fred and I were back on the road at five the following morning. It was dark and very cold, and the Gardner roared as Fred made her tramp on along the

A5. Pulling in at Markyate café we jenked up with derv, and because it was a full tank the pump attendant gave us money instead of cigarettes, as neither of us smoked. That paid for the teas and our *Daily Mirror*.

Most of the drivers in the café worked at the London Brick Company. They drove Leyland Octopuses and eight-wheel AECs. The drivers were a great bunch, always good for a laugh.

After leaving the café we made our way to the Merry Chest on the A2. 'We'll be there unofficially tonight, Shay,' Fred told me, 'because we'll be booking off at Barnet. So, do you still want to carry on working with me after this trip?'

'I wouldn't want to do anything else. How about me? Was I all right?'

'Not too bad, but you've got a lot to learn. I'll teach you all I know.'

'Thanks,' I said, chuffed.

'I'll fiddle the log sheets and expenses when I'm at home. What's the date today, Shay?' It was Saturday 11th March. 'Have a good weekend and I'll see you Monday morning, 9 o'clock sharp.' I walked away feeling very pleased with myself.

The following Monday morning we tipped at Strood then drove back to Banksy's yard. Deafy was going to do some repair jobs for us. The Governor was as nice as pie to me that day and I wondered if Fred had put in a good word for me.

While we were chatting in the garage, Deafy piped up, 'I'm having a new garage.'

'What's wrong with this one?' Sharpy shouted.

'What's wrong? What's wrong? Open your eyes, will you. It looks like a bloody colander, full of holes.

Anyway, Banksy told me I'm having gas heating.'

'You silly old bastard. He means you're having a bigger tortoise stove,' said Sharpy.

'I'm not taking any notice of you, Sharpy.'

'You will do this time next year. Why do you think he's building a new garage? Tell me that, clever dick. So that he can sell it more easily to British Road Services when they buy this load of crap. That only means one thing – you're out. You won't be creeping up Banksy's arse any more.'

'That's it! That's it!' Picking up a wrench, Deafy chased his tormentor out.

After a while Banksy walked in. 'Ain't I having a new garage, Governor?' Deafy asked him.

'I've already told you once. The company's having one built,' sighed Banksy.

'Sharpy said you're doing it so that BRS will buy us out next year.'

'How can they buy us out? They're not interested in us. We've got no contracts. We're a spot haulage firm.'

'Sharpy also said you've got a lot of A licences,' whinged Deafy.

'It doesn't make any difference to them, Deafy. It's the British government that's buying all the transport firms. They can print as many licences as they want. They call it nationalisation, but I call it gangsterism. I've told you before, don't take any notice of Sharpy. He's bomb happy.'

As Banksy left, Sharpy walked in. 'I've just asked the old man if I can borrow his car so we can all go out for lunch. Much to my surprise his answer was yes. So come on, lads, let's go!'

The five of us climbed into Banksy's new Wolseley –

what a squeeze it was. Fred looked at me and winked: 'Don't take too much notice of Banksy. He's rank Conservative.'

Sharpy drove up the lane like a bat out of hell. Three times we were practically airborne, the third time we landed badly. Reg reckoned the suspension had gone. 'It's making a terrible noise. You've wrecked it, Sharpy.'

'Bollocks to old Banksy,' he replied, and we all burst out laughing.

As we sat eating our lunch in a local café, Sharpy called out to the owner's wife, 'When you're ready, love, can we have some more teas over here?' As she bent over to put the tray down, he ran his hand up under her skirt.

'Don't do that, my husband's out in the kitchen. He's a very jealous man,' she cried.

'How about a date, then? I'll take you for a spin in my new Wolseley. You'd like that, you must admit.'

'I wouldn't mind.'

'We could have a rub out in the back seat. It'll make a lot of noise, though, because the suspension's gone.'

'How on earth did that happen?'

'Don't ask.' Sharpy didn't care a monkey's about anyone or anything.

Driving back, we all thought Sharpy would take it easy as the car was damaged, but not him – it was even worse this time. Sharpy really hated Banksy, and took it out on his car.

'If you don't slow up, boyo, we won't reach the yard. You Dunkirk soldiers are all the bloody same,' Dave Evans said.

When we got out of the car we realised that there

was definitely something wrong. Sharpy eased our consciences by saying, 'Don't worry, lads, it's got a year's guarantee.'

Just then Banksy walked over to Fred: 'I've got a load for you, out of West Malling aerodrome to a place called Over Kellet, not far from Carnforth.'

'Come on, Shay, let's get loaded. We're not coming back here tonight because Banksy will go berserk when he finds out about his new car. It's his own fault, though. He should either have given Sharpy a new lorry or sacked him years ago.' Fred grinned wryly. 'Deafy's made a good job of this windscreen wiper. He's a bloody good fitter. Served his time in engineering as a boy. Why he's wasting his time at Banksy's, I'll never know.'

We were now the proud owners of two windscreen wipers. 'If Atkinson Motors saw the wiper on my side,' laughed Fred, 'they'd certainly copy the pattern.' Normally they only had one.

On reporting to the guard-house, we were directed to a hard-standing, on which lay thousands of corrugated iron sheets that had been left there since the war.

An RAF feller driving a small Cole's crane came over to us and said in a very broad scouse accent: 'I've been told to load you eight hundred flat sheets and fifty curves. Is that right?'

'Yes,' Fred replied.

Let me know when you're full.'

After a while, Fred gave the thumbs-up sign to the driver who invited us for a cup of tea at the NAAFI.

'No thanks. We've a long way to drive,' Fred replied.

'Righto, lads.' And with that the airman was gone.

'Let's get out of here, Shay. We'll rope up outside,' said Fred. 'We've got about fifty buckshee ones onboard – if we get away with it we've got our beer money for tonight. I just hope they don't count them at the main gate. Apparently they're closing the aerodrome anyway.'

As we approached, Fred held his breath and sounded his horn. They waved us straight through. 'We've done it, Shay!'

We were anxious to avoid Banksy so parked up that night at the Merry Chest in Bean, near Swanscombe, from where we made our way home.

Chapter 3

The Fright of a Lifetime

THE following morning, I met Fred at six and within an hour we were driving through London, heading towards the A5. 'Fred, how long will it take to reach Over Kellett?' I asked him.

'Oh, about two and a half days. That's going downhill in the silent six with a tail wind,' he replied. 'You know about the silent six?'

'Yeah, Reg showed me on Swanscombe cutting. It's when you knock the gear lever from fifth into neutral.'

We drove on up Watling Street, doing roughly twenty-one miles an hour and eventually pulling in at the Markyate café for breakfast.

I suddenly realised on leaving the café that Fred had forgotten to pick up the *Daily Mirror*, so I ran back to fetch it. 'Where did you disappear to?' Fred exclaimed.

'You left the paper behind,' I said.

'That's right, blame me!'

'Well, there's no-one else, is there?'

'Saucy bugger. Come on, let's get cracking.' That's the way it was with Fred, always a bit of good-natured banter to start the day.

As we drove along, in the distance we noticed an Austin Seven car with a young woman standing beside it. When we got closer, it became clear that she was in trouble. There was not too much traffic about, so we pulled up behind her. Jumping down from the cab, we asked, 'Are you all right, Miss?'

'No, I've hit the kerb,' she said.

We found she had buckled the front wheel, damaging some of the spokes. 'Hit the kerb, woman? It looks as though you've been in collision with a tank,' said Fred in his usual to-the-point manner. 'Never mind, love. We'll soon get you back on the road. Shay, get the spare wheel out of the boot. Oh, and the brace.'

Walking round to the front of the car, I handed them over to Fred. He removed the nuts then, without a word, held the bumper in a vicelike grip and lifted the front of the car up off the ground. 'Don't just stand there gaping, take the bloody wheel off and put the spare one on – and be quick about it!'

'All right, all right, keep your hair on!' With that I put the three nuts on the wheel loosely, then Fred let the car down gently and tightened up. 'If I were you,' he told the young woman, 'I'd buy another spare as soon as possible. You never know what's around the corner.'

'Thanks very much. What do I owe you?' she asked. 'You've both been so kind.'

'Nothing, love,' said Fred smoothly. 'But if ever we meet in the future, which I very much doubt, you can buy us a drink. How's that?'

Thanking us again profusely, she drove off down the road.

I noticed that the young woman had been eyeing Fred most of the time. She seemed to be instantly infatuated with him, just like the rest of the women he met. 'You know what, Fred. You're a bloody idiot. Your luck was in there, and you didn't take the opportunity. Why?'

'Shay, it's okay if you want a one-night stand, which, I might add, I don't. Oh, and just for the record, remember this: if a woman will take her knickers off at the drop of a hat, she's not worth knowing. I'll know when I've met the right woman or not. Anyway, enough of this sort of talk, let's burn rubber. By the way, we'll be stopping at Cannock tonight.'

'Don't tell me. You know of a nice little pub with a piano.'

'I've told you before, Shay, sarcasm is the lowest form of wit, and it doesn't become you.'

'Sorry. I couldn't help myself.'

'The money comes in handy, though, doesn't it? It pays for our food, so when pay-day comes it can go straight into our bank accounts. As you are well aware, what we get in expenses wouldn't pay for a pot to piss in.'

As we drove past Tubbies on the A5, I gazed through the back window at all the curves in the air-raid shelters. The trailer was rocking to and fro gently; we probably had about seventeen tons on that day. The old Gardner always kept the same old rhythm, never missing a beat. It was like a hand-built clock.

I picked up the lavender polish and started cleaning the cab so that the timbers gleamed. The Atkinson had a beautiful coach-built cab and the more you lived with it the more it grew on you. The old engine was barking as we travelled along, doing about twenty miles per hour. Eventually we arrived at Cannock.

It had been hard going that day, with the rain coming down in torrents, flooding the road, which made driving to the lorry park very difficult indeed. The track was very narrow and going round bends was horrendous. Fred had to mount the kerb on several occasions and, believe me, an eight-wheeler is very heavy on the steering. Eventually we found a parking place where we stopped, checked the lorry as usual, and made our way towards the digs. By the time we arrived, our clothes were wet through, so we were very grateful for a hot bath and something to eat.

'Ready when you are, Shay. Let's go and have a pint and chat up the local talent. It's bound to be a bit quiet, though, being a Tuesday night.'

As soon as we walked inside the pub, you could cut the atmosphere with a knife, it was so silent. Everyone looked at us as if we had just arrived from Mars. 'It's like a morgue in here, Fred!' I complained.

'Don't worry. We won't be staying for long,' he promised.

Even the barman was not sociable, so we drank our beer quickly and left, not even bothering to say cheerio.

Eventually we found another pub, but virtually the same thing happened. When we walked in, everybody glared at us. All the same, we ordered a pint and sat down. Fred nudged me. 'Have you noticed anything? Everybody in this pub looks as though they've lost a pound and found a penny. What a shower! Funny looking lot, if you ask me.'

'Be quiet.' I giggled. 'They'll hear us.'

'I don't care, I'm fed up now. This isn't a very friendly place, is it? Drink up and let's get back to the digs. I've had enough of this. We'll have an early night.'

When we got outside, Fred said to me, 'Shay, I found it so embarrassing in there.'

'Why was that then?'

'I was the only one who was good looking.' When I laughed, he went on,'You think I'm joking? You must admit, Shay, one looked like Dracula, another looked like Frankenstein. They all looked brain-dead to me, especially the three in the corner. They reminded me of the three stooges.'

'You're a piss-taker, Fred, there's no two ways about it.'

The next morning we left early, and I complained to Fred that it was a bit nippy. 'Yes,' he agreed, 'it will take a while before the old Gardner warms up.'

Fred was really making her tramp on. 'Shay, keep your eyes open for the wooden-tops, they'll probably be lurking about somewhere.' The speedometer read around thirty miles per hour. Then I saw one of Sutton's drivers trying to overtake us with his eight-wheeler and trailer. 'Fred! Don't let him overtake!' I shouted.

'I can't go any faster, Shay. I'm flat out to the boards,' he replied.

The driver suddenly pulled out and drove alongside us, completely blocking the road. Both our lorries and trailers were fully loaded. I just sat tight and prayed nothing would come in the opposite direction. Sutton's trailer boy was rocking his fist up and down, giving us the toss-pot sign.

Looking over, Fred said to me, 'If they want a race, they've got one.' At that point the speedometer reached thirty-four miles per hour. The lorries were neck and neck; neither driver would give way. 'Shay, lift the bonnet,' Fred called to me. Pointing down to the pump, he yelled, 'Pull the bar back!'

'What bar?'

'That bar there!' he cried, pointing again. 'You stupid bastard, pull it back!'

In a short while the lorry gave more revs, and we gradually began to drive faster than Sutton's, who started to drop back. 'Fred, we're chucking out black smoke!' I said anxiously.

'That's nothing to worry about. I'm just clearing the shit out.'

I was still holding the bar back for all I was worth, and the road was flashing by beneath me. 'What speed are we doing, Fred?'

'Fifty miles an hour.'

'Fucking hell, we'll blow up!'

'Don't panic, it looks worse than it is. Shay, you can let the bar go now, and close the bonnet.'

'My ears are popping like crazy,' I moaned.

'Oh, don't make a fuss, you big baby,' he said, laughing. Looking out of the rear window again, I saw that Sutton's lorry had disappeared from view, leaving just a puff of black smoke.

'I didn't know you could make a lorry go like that,' Fred.

'Well, Shay the Gardner LW engine like the one we have is engineered to reach maximum safe revs at 34 mph (which as you know is 14 miles over the national 20 mph speed limit). To keep within the engine's safe limits a governor cuts in at 34 mph. A bar on the pump slides back in and cuts the revs out of the engine so you have no revs at all until you drop back down to 34 mph when the revs come back in. So, by pulling the bar back you are over-riding the governors and forcing the engine to give out more revs. Of course,' he finished with a wink, 'this is not something that should become habit. If you hold the bar back for too long you burn oil. Some drivers do it all the time, but end up putting in a gallon of oil a day.'

'Fred! I can see Sutton's lorry in the distance! He's gaining speed on us!' We looked out of the back window.

'Don't worry, Shay. He won't catch us now. We're too far ahead. And he definitely won't pass us. His lorry is chucking out black smoke, and that means only one thing – his bar is back. Quick, open the bonnet and pull that bloody bar of ours back again.'

Fred's foot was hard down on the accelerator, and our speed was nearing fifty miles per hour. 'Where are we now?' I asked.

'About five miles the other side of Stone, heading towards Newcastle-under-Lyme. I hope the old Bill isn't around, because we'll all get nicked, that's for sure.'

Both Gardners were still tramping on, billowing out black smoke. When we pulled in at Durose café, Sutton's parked alongside us. All four of us were laughing about our antics. Fred and I climbed down from the cab and joined the other two.

We introduced ourselves, and stood chatting for a while before going into the café. They seemed to be a decent couple of lads. I walked along with Bert, the trailer boy. Charlie and Fred were behind us. Once inside the café we all sat down, and Charlie bought the teas.

'How do you fancy swapping mates, Fred?' said Charlie with a grin.

'Why?' asked Fred.

'This one's hopeless.'

Then Bert piped up, 'No, we'll swap drivers instead!'

Charlie chuckled at that. 'We parked up at Northampton last night and are going back to our depot at St Helen's. How about you?'

Fred told him that Lancaster was our next stop.

After a while we all got up, said our goodbyes and were on our way.

We stayed behind Charlie and Bert all the way to Knutsford, then turned off and followed the signs to Manchester. We were now on the A6 heading towards Preston.

Suddenly I noticed something. 'Fred, there's smoke coming from one of the wheels!'

'Bloody hell!' he remarked, steering her over to the side of the road. We jumped down from the cab. 'Oh no! We've got a puncture on the second steerer, that's why I didn't feel anything on the steering. Good job you noticed it. Fuck me, it's red hot. You know what, Shay? You're a jinx.'

'I knew I'd get the blame for it.'

'Get the spare wheel from the carrier, Shay, and I'll start undoing the nuts.' As I was pushing the wheel to the front of the lorry, one of Pickford's drivers stopped. He had two mates with him, and they asked if we wanted a hand, which was much appreciated. They took it in turns to get the nuts off.

While we stood at the side of the road, two more drivers pulled up. They were from London, driving heavy haulage vehicles, and one of them shouted out, 'Need any more help?'

'Please,' called Fred. 'You know what they say: many hands make light work.'

'Is that so?' We all mucked in, and chatted as we worked. None of us were lost for words. It wasn't long before the job was done and we thanked them all. This was comradeship at its best.

Fred told me that when we reached Preston, he would book our digs in Lancaster, and we drove along, looking forward to the end of a hard day's work.

Above us the clouds were becoming ominously black. 'It looks as though a storm's brewing, Fred!' I had only just said this when there was an almighty bang, which made us both jump out of our skins. The lightning which followed was very vivid, then it started to rain, and in no time at all the whole road was flooded – the drains could not take the water fast enough. Fred shouted over to me, 'Are you all right? You're a bit quiet.' I didn't let on, but thunderstorms scared the shit out of me! There was so much spray that it was difficult to see, even though the windscreen wiper was working well. Deafy had made a good job of it.

When Fred finally stopped in Preston, the storm had subsided – much to my relief. He made a quick phone-call to book our digs and didn't stop again until we reached Lancaster.

In the car park I observed that a few Scottish drivers had already parked up for the night. Fred knocked on the door of the digs, and when it opened we were greeted by a very plump and jolly looking lady who made us feel most welcome.

Inside was comfortable and homely. The landlady

immediately put on the kettle and made tea for us, saying, 'I expect you two are hungry as well.' With that she came out carrying two huge dinners: home-made steak and kidney pies, potatoes, peas, carrots and cauliflower. She placed a gravy boat on the table so that we could help ourselves. As if this wasn't enough of a feast, it was followed by apple pie and custard.

When we'd finished, we thanked her, saying how much we'd enjoyed the meal. We chatted with the other drivers for a while, then one of them suggested a pint down at the local. 'I'll go along with that,' we all replied.

Down at the pub, one of Smillie's drivers was a great character who kept us all in stitches. After a while Fred walked over to the piano. He really liked playing, and he was so good that no-one ever complained. It was turning out to be a very enjoyable evening.

Soon some of the drivers started talking about how long they'd been on the road and how much experience they'd had, while in the corner of the pub stood a group of – how shall I put it? – a 'different class' of women.

Suddenly one of the women walked up to one of the Scottish guys and asked, 'Fancy a good time, dearie?' He upset her by telling her to clear off. I could see from the expressions on the local lads' faces that they were looking for trouble. A brawl was guaranteed.

Just as I was thinking this, fists started to fly. The more they fought the faster Fred seemed to play. The Scottish lads went berserk. Anybody who was within arm's length ended up with a black eye. A Smillie's driver opened the bar door and shouted, 'Throw the bastards out, and the women with them.' Blood was pouring from his nose, which looked like it was broken.

Five drivers rounded up a few local lads and threw them out the door, feet first. The landlord yelled, 'You can't throw all my locals out!'

'If you start, you'll be next,' bellowed one of the truckers. 'We'll soon find a driver to pull the pints.'

Fred looked at me, motioning that things were getting a bit out of hand. 'Let's make a move. See you later, lads.'

'The beer's free. Why are you going?' one of them asked us.

'Because we've got a perfectly good bed back at the digs, and we don't fancy sleeping in a cell tonight. I'd advise you lot to do the same.' With that we walked out the door.

There were still a few locals outside. Fred just pushed his way through, and none of them was man enough to take him on.

'Shay, there'll be no collection tonight, and if we carry on like this we'll soon be buying tea out of our own pockets. That'll never do. Never mind, we'll put up the price of the iron sheets.'

'Where will you sell them, Fred?'

'To a farmer if I can. Most of them are crooks – they're worse than truck drivers!'

The landlady greeted us as we walked in. 'You're back early, my dears. Had a good evening?'

'Not too bad, love,' Fred answered. He wasn't about to let on about the fiasco in the pub.

'I'll go and put on the kettle for a cuppa, I expect you could both do with one.'

'We won't say no to that. A cup of tea will do nicely, thanks.'

Next day, after tucking into a hearty breakfast, we proceeded to the car park, and promised Ma, as we'd come to call her, that we'd be back if ever we passed that way again.

But when we reached the lorry, there, sitting in a patrol car were a couple of wooden-tops, looking full of piss and importance. 'Oh no, what do those bastards want?' Fred muttered aloud.

'Were you two down the pub last night?' one of the coppers asked.

'No, we were at the pictures last night and didn't get in until late, officer,' replied Fred, all innocence. 'Anyway, my trailer boy is too young to go into pubs. What's wrong, then?'

'Oh, some Scottish lads took over the local and it got a bit out of hand. We just thought you drivers might have seen something. Well, thanks for your time.'

Bidding them goodbye, we jumped up into the Atkinson. Fred hit the starter button and we knocked the ratchet hand-brakes off. We drove out of Lancaster like there was no tomorrow.

We were soon back on the A6. 'We'll have to turn off soon and try to sell some of those sheets,' Fred told me.

We took the A683 and began to come across quite a few farms. Before long we saw a signpost for Wray. Fred suddenly turned into a very narrow lane. 'Bloody hell, Fred. I think you've dropped a clanger,' I said in dismay. 'You won't be able to turn round in this lane.'

'Who won't?' he replied.

In the distance I spotted another farm and Fred said this one looked likely. He parked up to have a word with the farmer, and a little while later they returned to see the sheets. I heard the farmer say, 'That's just what I could do with. How many did you say you've got? Fifty?' He looked puzzled. 'Couldn't you make it sixty? If the price is right, I'll do a deal with you.' I thought to myself, Fred's not going to sell ten more – they were supposed to be for somebody else!

'I'll tell you what,' said the farmer. 'I'll give you three pounds and ten shillings.'

'Make it four and you've got yourself a deal,' Fred replied.

'Okay, fetch the lorry in.' Very soon we had unloaded and stacked the sheets alongside the lorry. We walked back to the farmer's cottage where he gave Fred the money and his wife made us coffee. They were a very pleasant couple. I could see that, like most women, she had a bit of an eye for Fred.

They told us where our delivery was in Over Kellett and we set off. The road was almost the same width as us, but eventually we found our destination. We noticed they already had corrugated sheets there, and they put our load on top, not even bothering to count them – much to our relief. Fred looked at me and winked. 'We're off to Carnforth Haulage now, which is about half an hour's drive, to see if we can get another load for home.'

While we were driving, Fred handed me two pound notes. 'Thanks a lot,' I replied.

'Don't thank me, thank the RAF,' he winked.

'Fred? Would there be any chance of getting another load out of West Malling aerodrome?'

He looked at me with a mock old-fashioned air of disapproval, saying, 'That's the trouble with mates today, they get greedy.'

'Seriously though, I hope we do because the money does comes in handy,' I said, and once again Fred started to sing 'We're in the Money'. I wished he'd keep his hands on the steering wheel. It worried me at times.

At Carnforth we reported to the transport office. The governor took one look at us and said, 'You're a gift from heaven.'

'Why's that, then?' Fred asked.

'I've had two loads of paper delivered to Whitehaven but it's the wrong grade, and the reels are taking up valuable space. They want them cleared as soon as possible. They have to be delivered to the Imperial Paper Mills, Gravesend. I know it's out of your way, lads, but I'll pay top rates.'

I could see Fred's brain working overtime. 'Okay then, will do,' Fred replied.

'Thanks a bunch. Load what you can and I'll get another lorry to take the rest. I'll make it up with something else for Kent.'

When Fred and I were through we climbed up into the old Atkinson and headed for the Lake District.

By now the scenery was becoming rugged and hilly, with wide open spaces. On the approach to Kendal, we forked off onto the A591, making our way to Windermere. The road was very mountainous, water was running down into the gullies and the panoramic views were breathtaking. I would never have imagined this sort of setting in my own country. It was really beautiful and a complete contrast to what I was used to in Kent, the 'Garden of England'.

The roads were very narrow and in most parts there was just one lane. A lorry of our size was too big for this part of the country but then, only fifteen to twenty years earlier, lorries like ours had not existed. It had mainly been horses and carts. Fred had to move swiftly through the gearbox to get up those hills. Although the lorry was empty it was a laborious drag.

I asked Fred how long it would take us to reach Whitehaven. 'Well,' he said, 'we're at Windermere now. If we keep driving and don't stop off anywhere, we should be there in two hours.' The road seemed to be getting narrower and narrower. 'We're in Cumbria now. I love it here, Shay.'

Now and again lorries would come in the opposite direction, so Fred kept well over to the nearside. They were usually six- and four-wheelers, a lot smaller than our lorry. The drivers would give way and wave as they drove past.

Keswick was such a picturesque village, like something you only saw on a postcard. Fred turned left onto the A66, heading towards Cockermouth. The road was very similar to those Fred and I had travelled on before: there were the usual pull-ins to let traffic through, and drivers would normally give way to us because our lorry was larger than most.

The drive was very much like being on a roller-coaster, up hill and down dale. The bends were quite 'hairy' too. The roads were greasy as it had been raining heavily, and the sun had only just started to peep through from behind the billowy white clouds. It was turning into a pleasant day. The scent from the fields was redolent of spring, and I felt as happy as could be.

Signposts for Whitehaven were now in sight. 'We'll be there in about forty-five minutes, Shay.'

'Good. My arse is so sore. It feels as though I've been

sitting here for ages. I'll be glad when I can stretch my legs.'

'Oh, stop moaning.'

We reported to the gate-house, where they gave us our paperwork. One of the fellers directed us to a loading bay. Fred backed the trailer in, and they loaded us two reels high, putting wooden scotches in-between to stop them from moving about, and a backboard on the last one to stop them rolling off. We pulled out the trailer and backed the eight-wheeler in. While they loaded, we roped and sheeted the trailer.

Fred then drove the lorry out of the loading bay, backed it onto the trailer, and we roped and sheeted the lorry.

We stood and chatted to the men for a while. One of them mentioned a place where we could stay in Whitehaven. Fred thanked them, and we climbed back into the lorry. 'How's your arse now, Shay? Still attached to your legs, I hope.' Fred was grinning all over his face.

The next day would be a long one, so, Fred suggested a relaxing night in, which suited me fine. The following morning, after a good rest, we woke early and ate a hearty breakfast, washed down with two mugs of scalding hot tea. It never failed to set me up for the day.

Before setting off on our journey, Fred looked over at me and said, 'What's the date today, Shay?'

'Friday, 17th March.'

'Thanks. I'll make out a log sheet. That's always the first thing you do. Always remember that.' When it was done, Fred hit the starter button and pulled out of Whitehaven. 'We'll make for Warrington today, that's a good stop-off,' he said.

It was a long and arduous journey. Most of the time Fred had her in crawler gear, and whenever possible he put her into second, which made the double-drive shudder as it took up the strain. 'It looks a bloody long way down there, Fred!' I remarked.

'Yeah, we're about two hundred feet up.'

'It's a bit alarming, don't you think?'

'Well, when you've been on the road as long as me you get used to it,' he replied. Now and again the speedometer would touch twelve miles an hour and I could feel the heat coming from the engine. There were no two ways about it. We had the best diesel engine ever made.

On reaching the A66 our speed was just on sixteen miles per hour. I could feel the cold air blasting through the radiator grills.

The A66 to Cockermouth was a notorious road. Fred turned to me, saying, 'The brake drums must be getting hot.' When he braked you could hear the wheels whistling and the trailer wheels screaming.

We were descending rapidly down a very steep hill when all of a sudden Fred had to act quickly to negotiate a particularly sharp bend in the road. As he did so, he was suddenly confronted with a car travelling in the opposite direction but on our side of the road. It must have been doing about sixty miles an hour. I saw the surprise and horror on the other driver's face as he panicked and lost control. Fred swerved to the left, and I thought to myself, 'Oh no, he's going under the trailer that's for sure.' Fred pulled the steering wheel hard over to the right and the lorry leaned over. I just sat frozen in my seat.

Just as we came out of the bend, another was

approaching fast. Fred was fighting with the steering wheel to control the lorry. Our speed must have been at least fifty miles per hour at this point. I didn't dare even look. The smell from the brake linings was drifting up through the cab. Fred shouted over to me, 'Hit the brakes Shay!' By now my arse was really biting the buttons off the seat and it was difficult trying to get hold of the hand-brake, as Fred was driving on the nearside. He was running over small rocks, which seemed more like boulders to me. I finally managed to grab hold of the hand-brake and wind it up to its full capacity.

As Fred drove out of the bend, the lorry was still going down. The whistling had stopped, but all the wheels were screaming and the smell was obnoxious. The lorry began to shudder violently; Fred was having a hard time controlling it. Then, thankfully, the road began to straighten out a little. We started to slow up gradually and eventually we came to a halt, much to my relief.

'The brake drums are red hot,' Fred said grimly. 'If there had been another hill, the vacuum brakes would never have stopped us. I dread to think what would have happened.'

'Fred, I don't know how you managed to miss him, I thought you were going to run right over that car,' I said.

'Soppy bastard could have killed us!' Fred was shaken and understandably angry.

'I wonder how the car driver is.'

'Stuff him. He's probably lying in a ditch somewhere.'

We just sat there for a moment, recuperating in our

seats. I started trembling and thinking of the horror of it all. I'd had the fright of my life.

When we'd calmed down a bit, we climbed down from the cab to assess the damage. The heat from the wheels was quite intense. I knew the second steerers would be cold as there were no brake drums on eight-wheelers. As I was walking round the lorry I noticed something alarming about halfway along on the side. 'Fred!' I exclaimed. 'Look at this!' Two of the paper reels were sticking out by at least two feet.

'Fuck me!' Fred replied. 'That looks bad.' As he stepped over the drawbar of the trailer he noticed it was bent. 'That's put the front axle out of alignment on the trailer. Now I know what caused the terrible shuddering before we stopped. If I drive it like this I'll burn the tyres out in two miles. The tracking is completely messed up. We've got to get this drawbar repaired before we go any further.

'Let's have another quick look around to see if we can spot anything else wrong, Shay. When the drums cool off, I'll adjust the brakes. The main thing is the drawbar. The reels are no problem at all but we'll have to use an extra rope and double-dolly those two, then find a local engineering firm or garage where we can have the drawbar straightened out.'

He told me to get the hammer out of the toolbox, go to the other side of the trailer and pull the bar out of the four lugs. As I did so the tow-bar immediately fell to the ground. Fred knocked the split-pin out and removed it along with the washers.

'Help me unhook it, Shay, and we'll pull the complete drawbar away from it.'

'Fred! There's a lorry coming!' I yelled. Walking to

the side of the road, Fred put his hand up and hailed the driver, who pulled up right behind us.

'Is there a blacksmith in Cockermouth?' Fred asked him.

'I should think so,' the driver replied.

'Would you give us a lift, if it's not too much out of your way?'

'Yeah, no problem.' Fred dropped the sideboard of the other lorry down. We threw the drawbar on and away we all went.

Sitting in the middle of the man's cab, I thought to myself how strange it was looking out onto a bonnet. It was a Bull-nosed Bedford and, being petrol, it sailed along. 'I thought I'd had a poxy day,' remarked the driver, 'but in saying that, it's not half as bad as yours has been.'

In what seemed like no time at all, we were driving through the village of Cockermouth. A young woman was walking along pushing a pram, and the driver stopped and shouted to her, 'Can you tell me where there's a blacksmith or an engineering firm around here, love?'

'Yes, there's one just up the road on the left,' she replied.

When the driver pulled into the blacksmith's, Fred and I dropped the sideboard and pulled the bar off. It crashed to the ground. Fred looked at me, saying, 'Are you trying to completely balls it up?' and I laughed.

'Well, lads, I'll be on my way,' said the driver. 'I've a lot to do. Best of luck. Hope you get it fixed all right.' With that he drove out of the yard.

Fred knocked on the door of the blacksmith's and an old man appeared. His appearance was that of a proper country cousin: he was broad-built and his eyes seemed to be standing out of his head like organ stops. 'Can you help us out?' Fred asked. 'I've damaged the drawbar of our trailer.'

The old blacksmith eyed it up and down, asking, 'How on earth did you manage to bend it like that?' Fred explained to him about the accident, and what pressure the tow-bar had had to take.

'I see, I see,' the blacksmith replied, rubbing his chin. 'Fetch her in here and I'll take a look. I'll see what I can do, lads.' We watched as he used the bellows to heat the tow-bar and gradually turn it white hot.

Fred and I helped to lift the bar onto the anvil, then with a fourteen-pound hammer the blacksmith straightened it out with two mighty blows. 'It's as straight as a die now, but the only trouble is it will have a weak spot. I'll have to use my welding gear to reinforce it. That should solve the problem.'

The blacksmith cooled the drawbar down with water. He then pulled from the rack a large metal bar and cut off two twelve-inch lengths. 'This should do the trick because it is what I make my horse shoes with.'

He welded both sides of the damaged area with the two twelve-inch steel bars. With eyes rolling, he said 'that won't be a weak spot now, boy!

'How much do we owe you?' Fred asked. The blacksmith put his hand on his chin and, with rolling eyes, replied, 'Well, you've travelled a long way and you were in trouble. What about seven shillings and sixpence?'

Fred handed him a ten-shilling note, saying, 'Let's call it quits.'

'Oh, thanks. It's much appreciated,' the old man replied.

Fred and I lifted the tow-bar out to the path. 'How far away is your lorry?' the blacksmith asked.

'About three miles up the road.'

'Tell you what. I'll give you a lift if you like.'

'That's very generous of you. Thanks.'

'Right then. I'll go and fetch Ruby.'

'Ruby' turned out to be a 1933 Austin Ruby with spoked wheels. The blacksmith told us how he had converted it into a van himself. As we travelled along the road, he ground every gear in the box and left the hand-brake on until we told him about it. He was a good blacksmith, a clever man with iron and his heart was in the right place – but he would never have made a driver.

'That's our lorry over there on the right,' Fred told him. After five shunts, he turned the van around in the middle of the road to park in front of us. I was in pain with trying not to laugh, and when we got down from Ruby he was somewhat taken aback by the size of our vehicle. 'Dear, oh dear, isn't it long! How on earth do you manage to drive that?'

Fred looked at him a bit old-fashioned and said, 'How do you make horse-shoes?'

'That's easy,' he replied. Fred commented that so was driving a lorry. With his hand on his chin and with eyes rolling, the old man replied, 'Every man to his own trade, I suppose.'

Fred and I pulled the tow-bar from the back of the van and then Fred gave the old blacksmith an extra dollar. He was well chuffed. 'Well, lads, I wish you all the best.' He shook both our hands and was on his way.

Ruby bellowed out black smoke as he drove off up the road.

'Let's get cracking, Shay. The sooner we get this done the better I shall like it.'

'Okay, it's certainly been a long day.' We lined up the four lugs. Fred held the drawbar while I pushed the steel rod through them, then he put the washers on the end together with the split-pin. The next thing was coupling the tow-bar onto the lorry.

'Shay, if you get the jack out, I'll take the brakes up.'

As I turned the wheels, Fred adjusted all the brakes. I then removed the small granite stones from the road wheels. When it was all done, we put an extra rope on the trailer to secure the two protruding reels.

'Do you know what, Shay. We've lost three and a half hours over that stupid ponce who caused the accident,' Fred fumed.

'I know. It seems longer than that, though,' I replied.

'Get the expense sheet out, Shay, and book two pounds for the blacksmith.'

'Banksy will go bananas, won't he?'

'Fuck Banksy. He's out to catch us every day, but we only rob him once a month and the month's up. Book it!'

From the very beginning I had learned not to argue with Fred when he was in a temper, so I kept quiet.

'Since we're behind schedule, Shay, we won't reach Warrington until late.'

'It won't be the early bird catching the worm tomorrow, then,' I replied.

We both knocked the ratchet hand-brakes off and were on our way once more. I wasn't at all sorry to leave the outskirts of Cockermouth. After the trauma of the accident, the place didn't seem interesting any more. And when you've had a run of bad luck, nothing is ever the same.

I knew it was my imagination running riot, but travelling back along the A66 it now seemed more dangerous than when we had come. My body stiffened as Fred approached the bends, the memory of the accident fresh in my mind. I kept seeing the car and the look of horror on the driver's face as it came hurtling towards us. It had certainly scared the life out of me. Fred, on the other hand, took it in his stride – all in a day's work.

As Fred approached the bends he moved her down to fourth gear, doing all the fancy gear changes and tickling the accelerator pedal with his foot. He glanced across at me, giving me a wink, as if to say: 'Aren't I the clever boy!'

When we closed on the bends, Fred braked hard on the foot-brake while I kept the trailer in check with the hand-brake, and the next minute he was hard down on the accelerator pedal again, his feet going like a man playing the organ. But in all seriousness, when it came to driving an eight-wheeler and trailer, Fred was the best.

'I was thinking, Shay,' said Fred, 'as it's hard punch-driving through the Lake District and we've been delayed by the accident, I'll park up at Carnforth. We can have a good night's kip and get an early crack in the morning. That would make up the time we lost today.'

As Fred drove through Carnforth, we passed a cinema featuring John Wayne in *Stagecoach*.

'That picture's not that old! I've heard the drivers talking about it in the digs,' Fred commented. 'Shall we give it a try tonight?'

'I don't mind. I've not been to the dolly mixtures for a long time,' I replied.

We would book off at Warrington, he decided, and give old Banksy a good day's work. We'd make it up the next day, as it was Saturday and there wouldn't be much traffic about. We would be home Sunday night if we were lucky.

'Fred, I always wanted to work on lorries but I didn't think I'd be spending every minute of the day in one!' I joked.

As we sat in the cab chatting, all of a sudden there was a bang on Fred's door which made us both jump. Fred dropped the window down, and standing outside was a woman.

'Do you want to do any business?' she asked.

'No, I'm all right love, but I'll ask my mate.'

He turned to me: 'Do you want to do any business?' I was so shocked I couldn't speak.

Fred turned back to her and replied, 'He's not really sure, love.' She shouted to me, 'Five shillings for straight sex, half a crown for a blow-job and one and threepence for a hand shandy.'

Fred turned to me again and repeated what she'd said. I just sat there, not saying a word. I heard him tell her that I was saving up to buy a car, whereupon she kicked the front tyre and screamed, 'You can't shag a car!' Then she walked off in a huff, making her way to another driver who had just pulled in.

Fred closed the window, laughing. 'You know what, Shay, you slipped up there. She was beautiful and only about twenty-five. Still, it's your loss, not mine.' He was such a bloody piss-taker.

We booked in and had a meal. I hadn't realised I was so famished. It seemed a long time since we'd eaten.

'That's your trouble, Shay, you're always bloody hungry.' I denied it, pretending to be hurt, and he told me not to get shirty. Then he pulled a funny face, saying to me, 'Do you want to do any business?' In the end I couldn't help but laugh at him.

In the café, I noticed a driver sitting at one of the nearby tables, and I could see he was wondering what we were laughing at. Fred asked me if I wanted another cuppa before we made a move.

'Alright, then, you've twisted my arm.' As Fred passed the driver's table, he stopped and chatted to him, telling him about the woman in the car park. 'Do you want another?' he asked, pointing to the man's mug. Of course he accepted, as drivers always do!

After we'd talked for a while, the driver told us he worked for Cooper's Transport in Birmingham and that his name was Gordon. Fred asked if he fancied joining us at the pictures.

'I would love to but I'm a bit strapped for cash,' he replied.

'Oh don't worry about that. We played a good tune on the fiddle the other day and earned some extra money. So let's get changed and go and enjoy ourselves.'

The three of us got quite engrossed in the picture. It turned out to be a fair old film: John Wayne was jumping from horse to horse, and there was bags of action. We all thoroughly enjoyed it.

When we came out of the cinema it was dark and bitterly cold, brass monkey weather. As we walked along, Fred remarked, 'That driver's working late,' and when we looked, there was an eight-wheeler AEC fully loaded.

'He's working late but he's earning,' Fred continued. 'We'll all be back behind the wheel tomorrow. It doesn't matter what part of the country, or what time of day it is, there's always a lorry on the road somewhere, and it will always be the case – truckers north, truckers south.' I always remembered those few words.

I loved listening to Fred. I never tired of his stories. But if you upset him he was like the devil from hell. My attitude to life was getting more and more like Fred's. Since I was with him all day and every day, I suppose it was inevitable. The only thing I needed every day was a good meat pudding because my physique was not at all like his. Fred was all muscle, whereas I was what you might call lean.

We went for a jar, and after the pub closed and the three of us were walking back to the digs, Fred happened to look up. There in the sky he saw a shooting star speeding across. 'Look at that, Shay. If Banksy saw it he would say, "Chase that star. Make sure the rates are good. Load it, fetch it back, and we might earn a few bob."'

Carnforth was a pleasant small town, very clean and respectable. Most of the houses had white net curtains hanging up in the windows, and some even had boxes on the sills with plants in them, ready for the spring. The people we met were friendly. That was the thing about road haulage: you saw a lot of new places and people. If it wasn't for that job I wouldn't have seen Carnforth.

When we arrived back at the digs, the landlady made us a pot of tea. 'Do you think you could give us all a call in the morning?' Fred asked her.

'Certainly. What time? Would four-thirty be too early?'

Four-thirty was fine, and the next morning, with a good breakfast inside us, we all left the digs at about half-past five, said our goodbyes and were on the road by six-thirty. 'It will be a long old day, but we'll soon make up for what we lost yesterday,' Fred promised me.

Once we were on the A6 our speed soon reached thirty miles an hour. 'Pull the bar back, Shay.' Moving the blankets off the engine bonnet, I lifted the cover. The Gardner was roaring heavily. Seeing the road passing under the axle quickly, I pulled the bar back while Fred kept her at forty miles per hour and I kept my eyes peeled for wooden-tops.

The A6 was a bumpy road, as the tarmac was very uneven. One minute the camber was down, the next up, and the lorry was rocking from side to side. It took all my might to stay seated. Drivers called this notorious stretch the 'dart board' because it had become so pitted over the years. Now and again Fred would drive over a pothole and the shock of it would pass through the steering column and knock his hands off the steering wheel. 'Shay - keep your eyes on those poking-out reels. We don't want to lose them.'

'Don't worry, I'm watching out.'

By now we were coming up to Manchester. Fred told me to let the bar go and then to pull it back again once we'd passed through the town. Fred never let up, the black smoke billowing behind us. It was the first time I'd known Fred to push his lorry to the limit like that. We seemed to be flying – over railway crossings, telegraph poles flashing past. Fred just tramped on. I must admit I was enjoying the ride. 'It's a bit windy, Fred! It looks like rain too.'

I had only just mentioned the rain when the heavens literally opened up and down it came in torrents, flooding the road. 'That will cover up the trail of black smoke we're leaving behind,' Fred remarked. 'And at least the wooden-tops won't be lurking about.'

Fred was now driving through villages again. The wind was howling and hail was pelting down, which made the roads treacherous. It was almost impossible to see through the windscreen. Fred stopped only just in time at a zebra crossing where a young woman with a pram was waiting to cross. She put her hand up to thank him as she passed in front of the lorry.

The rain began to ease a little and the sun was trying its best to break through. It was still very windy. 'Not long now, Shay. We've done well and made up the time. Thank goodness we didn't lose those reels. We would have been in trouble.'

As we passed Monks Heath on the A34, Fred was gaining fast on a four-wheel Foden. It had started to pour with rain again and visibility was bad. Before I knew what was happening, Fred had pulled out and was overtaking it, his boot hard down on the accelerator. As he pulled the steering wheel hard over to the right, instinctively I pressed down hard with my foot on an imaginary pedal, trying to make the lorry and trailer go faster, praying at the same time that nothing would come in the other direction, as there was no overtaking on this road. Once Fred had passed the truck, I sighed with relief. The driver flashed Fred in. 'Thank Christ for that, Fred. You really put the shits up me at times.'

'When I said we had to get our finger out, I meant it,' he said, looking over at me and grinning saucily.

By the look of the sky, the weather was not going to let up. Soon Fred started to slow down as we came into Newcastle-under-Lyme. Once there he didn't hang about. He did a quick gear change but kept up the same momentum all the time as we travelled mile after mile.

As we approached the traffic lights in Stafford, they changed to red. In an instant I hit the ratchet hand-brake, coming to a halt two feet past the traffic lights. Fred glanced across at me. 'You did well there, Shay.' No sooner had the lights changed to green than Fred sped away as fast as he could. I could feel the strain of the double-drives as he did so.

Within forty-five minutes we were on the A5. As we passed through Cannock, Fred didn't let up but drove fast and furious, pushing the lorry to her limits.

'Let the bar go now, Shay,' Fred shouted as we began to go down a hill. As I did so, Fred knocked the gear lever into the silent six, and away we went. I don't know how he felt, but I was feeling knackered and hungry.

As the road started to straighten, Fred put the gear lever back into fifth, hitting the gas pedal hard which made the engine rev more. All of a sudden Fred had to ease up, as there in front of us was a steam lorry doing only about twenty miles an hour. 'I can't stay behind him, Shay. It'll drive me up the wall.'

After driving behind the steam lorry for about ten minutes, Fred pulled out and overtook him. The driver blew his whistle and waved as we passed. 'Shay! I bet he's calling me a mad bastard, passing him like that, because I'm well over the speed limit for this road.'

On the A5 the traffic was heavy. I commented that the lorry in front of us was a nice one. 'Yeah, it's an eight-wheeled Maudslay. They're not bad lorries if you don't want reliability,' he replied, putting his foot down again and overtaking it. 'In this modern transport world, Shay, there are three good things that you need: an ERF or Atkinson, with a David Brown gearbox and a Gardner engine. They just keep going and don't seem to wear out.'

Eventually we pulled in at Tubbies café. When I climbed down from the cab, I was bow-legged like Gabby Hayes from one of the cowboy films. 'Don't walk like that, Shay. It looks as though you couldn't wait.'

'What do you expect, sitting in a lorry for eight hours?'

'You lads don't know you're born,' Fred replied, laughing.

Sitting down at last, we tucked into a very welcome dinner. Fred said he had booked on at Warrington, and booked off at Towcester for that night. We had done what you would call a good day's work. As the next day was Sunday, Fred asked me what I'd rather do: drive through the night or park up?

'I'll fall in with whatever you say, Fred.'

'Right then, I'll derv up, check the lorry over and as you've held the bar back a few times it will probably need engine oil. We'll have a wash and brush-up here, have a nut-down in the cab, then make our way home on a dodgy night out. Have you got that, Shay? We'll book off at Towcester but we'll be parked up in Gravesend.'

I seemed to have been asleep for only ten minutes when suddenly I was woken by Fred starting the engine.

In no time at all we were back on the A5. There were a lot of other truck drivers on the road. They came from all over the country, all making their way to London. The night-drivers really made their lorries sing.

The roads in Towcester were very narrow, and the noise from the engines must have been terrible for the occupants of the houses along the street. They must have felt their buildings shudder when the truckers drove past, nearly every one of them breaking the law.

At the Markyate café the lorry park was jam-packed, leaving us just one space. It was a change-over place for night trunkers, and day drivers slept there for the night. We chatted to some of the other drivers while we had our tea. When they left there were only Fred and I sitting in the corner of the café.

I asked Fred what his wife had to say about him staying away from home so much, and he told me he was divorced. Embarrassed, I apologised.

'Well it's been quite a while now,' he confided. 'I was working for Pickford's Heavy Haulage before and during the war, driving virtually day and night. We didn't have log sheets in those days because we contracted for the Ministry of Defence. Whatever we did had top priority – we were carrying things like tanks and guns. You name it, we carried it, delivering mainly to the docks.

'My wife was a career woman, and we hadn't been married long when I discovered she was having an affair. I just walked out and left everything to her. I've never laid eyes on her since.'

I remarked that it all sounded very sad, but he said, 'No, Shay, when I think about it, we weren't really suited. That's just the way life goes sometimes. Anyway, I left my job so that she wouldn't know where I was. But in leaving Pickford's I cut off my nose to spite my face, which was a bit silly, I know.'

He'd met Banksy when Banksy was driving a six-wheeler Leyland doing Ministry of Defence work. Fred had been on the main road at Enfield when he noticed one of Banksy's lorries coming through and hitched a lift. Banksy had just bought an eight-wheel Atkinson and trailer which the lads there at the time wouldn't drive because of its size and also because it was journey work. 'So that's how I ended up being employed at Dartford and working for Banksy. I have digs in Brent Road, Dartford, which are temporary, so all my correspondence goes to the office. That's where you're lucky, Shay, living with your mum and dad. Make the most of it, lad.'

I commented that with Fred's looks he wouldn't have any trouble finding a woman. 'No way! I'm not making the same mistake twice. But I'll tell you this: by long hours and hard work, I can put money in the bank and my advice to you, Shay, for what it's worth, is to do the same. As well you know from the work you've done with me, there's no such thing as a short day. I'll always fiddle it so we get at least eleven hours. Anyway, that's enough of me boring you to death with my life history – let's make our way home on a dodgy.'

So on we went to the A2 where we parked up at the Jubilee café and hitchhiked home.

The next day we tipped our load at Gravesend, then made our way back to the depot. When we pulled up to the derv pump, I noticed a lot of activity going on. Reg walked over to us and climbed up onto Fred's side of the cab, saying, 'Banksy's going loopy about his car. He's blaming us all. He'll probably tackle you about it when you see him.'

Fred told him not to worry, he had no intention of saying anything. He didn't want to get involved in that. 'Anyway,' Fred asked, 'what's going on here, Reg?'

'They're building a new garage for Deafy.'

Meanwhile, I jumped down from the cab while Fred went to the office. The old man greeted Fred warmly, congratulating him on making a good profit for the company that week.

'Oh, by the way, Gov, I had a slight accident and bent the trailer bar. Before you have a go at me, I've had it repaired, but it's going to cost you a few bob.'

'Don't you worry about that, Fred. We haven't got much work on, so get your lorry serviced. Deafy will go right through it for you.

'Reg is on a week's holiday so you can drive his lorry on Truman's Beer. Shay can always help Deafy out if there's no other work in the offing.' So it was all sorted out and the old man turned to go and work out our wages and expenses.

Then he remembered something. 'By the way, Fred, when you were in my car going to the café, how did it get damaged?'

'Damaged?' said Fred, all wide-eyed surprise.

'Yes, you heard me! Don't play the innocent with me.'

'I don't know anything about any damage, Gov.'

'Well, somebody's not telling the truth. The chassis doesn't just break in half on its own. It's the last time

I'm loaning my car to anyone. More sodding money I've got to pay out. It's going to cost me a fortune.' Still Fred played the part, swearing the car was perfectly okay when we brought it back.

When I was told what was happening, I was sick inside. Fred put his hand on my shoulder and told me not to worry. Suddenly Reg shouted out that the governor wanted to see me. 'You don't know anything about the car. Okay?'

I knocked on Banksy's door, and he called for me to come in. He began, 'Well, since you've been here, I've been very pleased with your progress. I am going to give you a pay rise of two shillings and sixpence.' I was shocked but delighted, and I thanked him.

'Everybody likes you here, Shay,' Banksy went on. 'You are an honest, genuine and a very bright young man.' Then he changed tracks abruptly: 'By the way, do you know anything about my car?'

'No, Sir. What do you mean?'

'When you went up to the café, what actually happened?'

I thought to myself: crafty old bastard. I told Mr Banks how Sharpy had driven carefully up the road in his new car, and done exactly the same on the way back. 'Everything seemed perfectly all right to me,' I said.

He gave me a puzzled look: 'Shay, you'd make a good driver, because you are like the others, a blasted liar. I take back what I said about you being honest. Now get out of my office.'

As I walked out, I turned round and said cheekily, 'Do I still get my pay rise, Sir?' Then I wondered immediately afterwards how I'd dared, and felt myself go very red.

'Yes, you've still got your pay rise,' the governor replied, smiling. 'Now scram.'

The old man informed me that I would be working with Sharpy the following day. I must admit I was a little apprehensive about it, wondering what it would be like after being with Fred.

When I met Sharpy early in the morning, he told me we had to deliver six tons of flour to Canterbury. The lorry was already loaded, so we were soon on our way.

As we drove up Chatham hill the noise from the Perkins engine was deafening, and the heat intense. The road evened out as we headed towards Rainham, and I bellowed to Sharpy, 'How do you put up with this noise?'

'The old Vulcan's clapped out. It's had its day,' Sharpy yelled back. 'The governor told me last week that I'm going to have a new four-wheeled AEC. If I don't get it, Shay, he can stick this lorry right up his jacksy, no messing.'

'You don't like Banksy, do you?'

'No, because he made his money during the war, doing a bit of wheeling and dealing, while we poor bastards fought it and earned fuck all. I've told him that to his face. He was a much better person when he didn't have a pot to piss in. Money seems to have gone to his head. We both worked for Barton's timber years ago, so I know all about Banksy. He fiddled more timber than he delivered.'

Sharpy wanted to get to Canterbury as quickly as possible to get the stuff unloaded because, if it was all right with me, he said, he wanted to visit his sister. I said that was fine and he drove like there was no tomorrow. When we arrived, the pair of us unloaded the flour. It was very hard work that made my arms ache. Afterwards, we used their facilities to get cleaned up. 'Okay, Shay, now I'll take you to visit my sister, Rosie,' grinned Sharpy.

We parked up in one of the back streets and walked to St Peter's Place. Sharpy stopped outside a house. 'This is it, Shay.'

The neighbour was cleaning her windows. 'Hello,' she called out. 'Come to visit your sister?' Sharpy nodded as he rang the doorbell.

The door was opened by a woman of around thirty, not bad looking either. Before she had even closed the door, Sharpy was kissing her passionately, saying, 'Oh it's so good to see you again, Rosie. It's been such a long time.' I thought to myself: he's forgotten all about me. Then suddenly he realised, apologised, and introduced me to her.

'Friends of Sharpy's are always welcome here,' Rosie said with a wink, and I felt myself blush with embarrassment.

'Nice to meet you,' I stammered, feeling very foolish. Rosie showed us into the sitting-room, which was very clean and tidy.

Sharpy asked: 'You don't mind amusing yourself for a while, do you, Shay? Only I've got some business to attend to.' He was winking at me.

'Of course not,' I said, not knowing what else to say.

As Sharpy walked with Rosie to the bedroom, he glanced back at me and grinned. 'I expect you've already gathered that Rosie's not really my sister. Her husband was a friend of mine in the army. I promised I'd take care of her if anything happened to him. He copped it in the war.'

As I sat in the front room, I could hear noises from the bedroom. The walls were paper thin and I tried to imagine what sex was like. I had no idea. I heard Rosie giggle and call Sharpy a dirty old bastard. Goodness knows what her neighbours thought, since they were supposedly brother and sister.

Eventually the bedroom door opened and Sharpy appeared, naked, his clothes tucked under his arm. 'There is nothing on this earth like a good shag,' he declared. 'By the way, Rosie says she fancies you, Shay.'

'Who, me?' I was totally taken aback.

'Yes you – so get in there and make hay while the sun shines. It'll make you feel bloody marvellous.'

I heard Rosie calling to me, but I just stood transfixed, unable to move for a minute. The next thing I felt was a hand on my back. It was Sharpy pushing me into the bedroom.

Rosie was lying on the bed wearing just a flimsy, see-through negligee. I sat down gingerly on the edge, feeling very randy but at the same time nervous.

Rosie got up and started undoing the buttons of my shirt, gradually sliding it over my shoulders. It fell onto the bed. I noticed her hands and fingers were very slender. 'Stand up, Shay,' she commanded. When I stood up, she undid my zipper and belt; my trousers fell to the floor.

By this time I was perspiring profusely. I quickly undid the laces of my shoes and slipped them off, and as I stood up again, quick as a flash my pants were down. Rosie lay on her back, pulling me on top of her. The feeling was unbelievable. She was warm, juicy and tight.

'You were a virgin, Shay. Am I right?' Rosie asked afterwards in her gentle voice.

'Yes, you're right,' I answered.

I felt absolutely fantastic. Eventually I recovered enough to leave the bed and get washed and dressed.

Rosie made us a pot of tea and a few sandwiches. 'I live on my own,' she said, 'and get lonely. Sharpy visits me whenever he can.'

Then later on, as we were walking to the front door, Rosie whispered, 'You're always welcome to visit as well, Shay.'

She stood on the path and waved as Sharpy and I walked up the road.

'What did you think of our Rosie, Shay?'

'She's certainly a lovely woman, good looking an' all,' I answered.

'Shay, what went on today is strictly between ourselves. All right?' Sharpy said.

'Mum's the word,' I promised.

We were now on the A2 heading towards Folkestone, and Sharpy told me he would drop me off at Gravesend on the way home, if that was okay with me. He reminded me not to forget to ring the office to see what time I was starting in the morning, and we said our goodbyes.

As I walked along the road, my mind went back to my afternoon with Rosie. Just the thought of it made me feel randy again, so I pushed the thought away. Sharpy was no transport man, but he was certainly a ladies' man. He had no scruples whatsoever, but I just put that down to his experiences in the war. As the governor said, he was bomb happy.

I started work at eight the following morning. The

governor told me that I would be working with Deafy. As I walked into the garage, Deafy came hobbling over, chuckling like an old mother hen. Two of his fitters were with him, and as they drew closer Deafy shouted, 'Oh, the Canterbury stud's arrived!' At first I didn't cotton on to what he was saying, and then the other two fitters started.

'Who had his leg over, then?'

'We hear you were at it for two and a half hours, randy little git!'

Then Deafy piped up excitedly, 'Tell us more, Shay. Tell us more.'

I could not believe what I was hearing. Sharpy had distinctly told me not to say anything, yet he had done quite the opposite – told a true story but exaggerated and dropped me right in it. Bastard, I thought. 'Shut up, Deafy. I don't want to discuss it with you or anyone,' I growled.

'Ooh, we are ratty this morning. Got out the wrong side of the bed, did we?' Deafy mocked.

'No. I'll tell you all about it when we break for tea. All right?' I knew they would hold me to that. When something like this happened, especially in the truckers' trade, you never lived it down. Your workmates would take the piss out of you mercilessly. They just loved it.

Deafy gave me the job of painting wheels that day while he made a new tow-bar for the trailer. It was really boring. The fitters stripped Fred's lorry and trailer down, and I must admit they were good at their job.

Deafy called out, 'Shay, want a break from that?'

'I wouldn't mind. This job's a pain in the butt.'

'I know but it has to be done,' he replied. 'Walk over to the office and get some writing paper for me.'

As I approached the office, Kay, Banksy's secretary, looked up and smiled at me. 'Who had his wicked way, then?' I couldn't believe what I was hearing. Even the secretary knew!

The governor inadvertently overheard Kay. 'What's this I hear about you getting carried away and ravishing a woman, eh? Dirty lucky sod!' he continued with a look of amusement on his face. 'But if you think I'm paying you to have leg-overs on the job, you are very much mistaken.'

I was in such a temper that I snatched the writing paper off Kay and walked out, thinking to myself: Dad's an eight-wheeler driver, he's sure to hear about all this. I'll wring Sharpy's neck when I see him.

I was in the garage when Reg cycled in. 'Hello Shay. Busy painting wheels, I see. You're doing a grand job.'

'Thanks, but I'd much rather be with you,' I replied.

'I heard you were in the garage today. How would you like to come round for some lunch? The old girl would love to meet you.'

'Well that's good of you, Reg. I'd love to.'

'Right, then. I'll go and have a quick word with Deafy to see if I can persuade him to let you off for a couple of hours. Perhaps he'll loan you his push-bike, as it's a bit of a hike to get to me otherwise.'

Reg seemed to be with Deafy for quite some time. When he finally came out of the office, he shouted over that Deafy had said I could go early.

On our tea-break the lads kept teasing me about my shenanigans with Rosie, and I could feel myself getting uptight. 'You'll be working with me again tomorrow, Shay,' Deafy commented, 'so you might as well tell us about your lustful night on the tiles.' I decided at that moment nothing would make me tell.

'The old man's gone out, Shay, so you can nip off around half-eleven. Be back by three, though, in case the governor decides to come back. You never know with him. Oh, and about my bike – take it, but make sure you bring it back in one piece, and tomorrow you can tell me all about your sexual encounters.'

Half-eleven came and I was gone like a flash. When I arrived at Reg's house in Vincent Road, the time was quarter to one.

Reg introduced me to his wife, Elsie, who was a little shorter than Reg. The first thing I noticed was the dimples in her cheeks when she spoke – a very attractive looking lady, I thought. 'Call me mum, Shay,' she told me. 'I'd like that.' The ploughman's lunch she'd prepared was really tasty and there was plenty of it: home-made pickle, and rolls she had baked herself. Elsie made me feel very welcome, though she must have thought I looked half-starved because she kept offering me more food. Eventually I had the willpower to say, 'No more, please.'

While Elsie was washing up, Reg and I went out to have a chat in the garden, which was very neat and well kept, with a lawn with flower borders around it, a few shrubs and a rockery in the corner. It was what I would call manageable. Around us the birds were singing their little hearts out. Reg lit his pipe. 'You're chucking out more smoke than a Vulcan,' I told him. He smiled.

We sat down on a bench and I remarked how pleasant it was. 'We like it here, Shay. It does us,' he replied.

Deafy had already primed Reg, so I told him everything that had happened between Rosie and me, sparing him the details of course, and how Sharpy had stated most emphatically that he would not mention it to anyone.

'Shay, listen to me. Don't take it to heart. No harm's meant. They're a good bunch of lads. You'll get ribbed, but my advice, for what it's worth, is don't bite. This will die a natural death after a while, you'll see. I know you want to have a go at Sharpy, but forget it. Turn the cards on him by agreeing with everything he says. When he's there tell everyone that Rosie said she prefers you because you're younger and more nimble. That'll soon quieten him down. But believe me, Shay, Sharpy is all right.'

Reg also told me to remember that I hadn't done anything wrong, and that the ribbing was all part and parcel of growing up.

Then we saw Elsie coming down the garden with three glasses of lemonade on a tray. 'Let the Missus make a fuss of you,' Reg whispered. We lost our boy during the war. She misses him so much. We both do.'

Elsie put the tray down on a small table. 'We're going for a holiday to Jersey in a couple of days, Shay. Did Reg tell you? I've never been there before. Do they speak English over there? Only, the old man here says they don't. He teases me something shocking at times.'

I smiled at her, saying, 'Yes they all speak English.'

Elsie told me she had heard all about me from Fred Ruddock.

'Fred!'

'Oh yes, son, we see Fred often. He comes round for dinner and a chat. He's a lovely man, handsome with it,' Elsie said with a cheeky grin, showing off her dimples. I told her how much I liked Fred too, and couldn't wait to get back on the road with him on Monday.

'How long have you been with Banksy now, Shay?' Reg asked. 'It's been about 3 months now, and the drivers seem to have taken to me, even old Banksy.'

'It makes working much more pleasant if you like what you're doing and get on with the people,' said Reg.

Reg and Elsie made the afternoon really enjoyable for me. It made a nice change, though it went far too quickly. The advice Reg gave me about not taking life and people too seriously made me think. I was learning and growing up fast. Time had passed so quickly with Banksy and I loved every minute of it, working on heavy lorries.

It was now the winter of 1947. I remember it as though it was yesterday, because the winds were really biting. We had to endure all weathers: hail, snow, fog and ice. The water in the engine nearly froze, even the diesel at times. It was one of the most horrendous and hazardous winters in our history.

Chapter 4

The Lancashire Lass

THE drivers still teased me, but I answered them back now, giving as good as I got, which surprised them. It was all good fun. I wouldn't have done that a few months earlier. I felt much more confident in myself because of what Reg had said to me.

Fred and I had covered a lot of miles together. Although it was hard work, we had plenty of laughs, and Fred had become not only a working partner but a friend as well.

One day in February 1947, we had just loaded Gyprock plaster board out of Rochester. The first night, we stopped at Brownhill's, the Jubilee café on the A5. The weather was really bad, and my whole body felt numb with the cold. I was wearing a thick overcoat which had been given to me. There was a couple of buttons missing, so I held it tight around me, putting the collar up as I desperately tried to keep warm. That night Fred drained the water from the radiator, poured it into a ten-gallon drum and placed it inside the cab on the passenger side, covering it with an old sack to try and stop it from freezing.

As we left the digs that morning, our surroundings looked very picturesque – everywhere was covered with snow. While I scraped the ice and snow off the windscreen and windows, Fred took the drum out of the cab and filled the radiator, thanking his lucky stars that it had not frozen during the night. We shivered, and our noses kept running. That morning there were about ten drivers out on the lorry park who all mucked in and helped each other. Some of the lorries refused to start. As drivers pressed their starter buttons, someone would turn the starting handle at the same time to help turn the engine over. One Seddon just would not start. I heard Fred blaspheme, calling it a Catholic bastard. 'What are you, Fred, Church of England?' I asked him. Looking at me with knitted brows he answered, 'No, Shay. I'm a Catholic.' I just wished the ground would open up and swallow me.

Fred found a piece of wire, tied a piece of rag to it, dipped it in the diesel tank and told the driver to lift up the engine bonnet inside the cab. He then pulled off the air breather, lit the rag, and as one of the other drivers pulled on the starting handle, the driver of the lorry pressed the starter. Immediately Fred stuffed the naked flame down the breathing tube, and she burst into life, coughing and spluttering, chucking out more shit than an elephant in a zoo. Fred jumped down from the cab with the flame still burning high on the wire in his hand, walked around to the driver's side and shouted, 'Driver! I'll do you a favour, mate. You jump out, and I'll chuck this lot into your cab.' We all laughed.

'Well, lads, let's go and have a cuppa. It'll warm our cockles before we set off,' someone suggested. We left the lorries ticking over merrily on the lorry park.

We followed each other onto the A5, heading north. As we drove through Newcastle-under-Lyme, my feet and fingers went numb again with the cold. I told Fred

I didn't know how he could drive with it being so cold. 'I've been doing it for years, Shay. You just have to get used to it.'

Suddenly Fred pulled up right in the middle of the road, blocking it completely. He could not pull over to the side because there was too much snow. 'I've had enough of this, Shay. I'm covering the radiator with that old sack, and putting rag down by the hand-brake to stop the draughts coming through. Maybe it'll make us feel a bit warmer.'

The other drivers waited patiently behind us, and within a couple of minutes Fred pulled off. I could feel the double drive wheels spinning as we gathered momentum. After a couple of miles, the cab started to warm up a bit – instead of being freezing it was just bloody cold! 'I'm a bit worried, Shay. I shouldn't have covered the radiator up like that. It's dangerous. It could damage old Betsy. As soon as I find somewhere to pull over I'll jump out and drop the sack down a little.'

I asked him how it could damage the engine, and he explained that if the engine got too hot it could damage the liners, gasket and head. That was why you could run a lorry low on oil, but not on water. 'But you haven't got to worry your pretty little head about any of that yet, because I'm the captain of this ship and whatever I say goes.'

The windows were icing over inside the cab. I tried to keep Fred's side as clear as possible so that he was able to see better. The weather was becoming worse by the minute. Then all of a sudden, coming in the opposite direction, we saw an old boy driving a five-ton Thornycroft with no windscreen or windows;

all he had was a small piece of canvas, waist level. He was sitting behind a steering wheel that was made of wood and with what must have been a king-sized horn on the side. The driver looked just like a mummy, he was so well wrapped up. Neither of us could believe our eyes. 'Fred! I'll never complain about the weather again after seeing that poor bastard. It makes me feel quite guilty. I'm glad I'm not his mate.'

'Oh, he's one of the old truck drivers. What you might call a knight of the road.'

It started snowing heavily again, and the windscreen wipers could not clear the snow away fast enough. Every now and again Fred would put his hand out of the window and try to clear it. It was terrible. I felt really sorry for him. His eyes were streaming, not only from the cold but with straining them trying to see out. 'The Knight of the Road doesn't have this problem,' Fred commented. I laughed.

Now and again the eight-wheeler would slide on the ice. The convexity of the road was awful. The trailer didn't help either. It was attempting to push the back end of the lorry round, making it really dicey. I remarked that the only trouble with being a transport man was you had to keep proving it by driving in all conditions. 'Not really, Shay. You need a lot of bottle, that's all.'

It was now 11.30 in the morning. Fred had not driven at more than ten miles an hour since leaving the Jubilee café. It was ridiculous driving in these conditions. Fred asked if I thought it was strange that we hadn't seen many cars coming the other way, or even lorries, come to that. 'I wonder if there's a problem ahead,' he said. 'I can't do anything about it now even if there is. The lads are still behind us, Shay, so if there's anything wrong they're going to cop it too. We'll just keep trunking, as they say.'

'I don't know about trunking, Fred, it's more like trudging the speed we're doing.'

As we neared Mere on the A50, the weather wasn't getting any better. It was like an ice rink, and the lorries behind were not having much more luck than we were. I said I thought it only rained in Manchester. 'I wish it would turn to rain. It would make driving a lot easier than this snow,' Fred replied.

In front of us now we saw a police car, with an officer standing beside it. 'That's all we need,' Fred commented.

After flagging us down, the officer walked round to Fred's side, looked up at him and said, 'Where do you two comedians think you're off to?'

Keeping his tongue between his teeth, Fred replied that we were parking up in Lancaster. 'That's what you think,' the officer replied. 'You're not going anywhere. The road's closed ahead. Didn't either of you see the sign saying road closed?'

'No, officer, we didn't see anything.'

'Of course you didn't. You probably ran over it and flattened it,' the officer replied sarcastically. 'Sorry, lads, you'll have to park up in Warrington, and find yourselves some digs for the night. Ring this number and let them know where you're staying. They'll phone you when you can set off again.'

By now all the other drivers had gathered round. 'What's the hold-up?' one of them shouted. 'The road's closed for some reason. They didn't say what was wrong, but we've got to park up in Warrington. So, sort yourselves out, lads, and get organised.'

Fred whispered to me, 'Let's get cracking, Shay. Otherwise the places will fill up.' As Fred drove off, the tyres began to grip the icy road more easily. It had started to thaw.

We were now on the A50, heading towards Warrington, about fifteen miles away. When we reached our destination, the police stopped us again and told Fred to pull up behind the last truck that was parked alongside the road. There must have been about twenty-eight lorries in total.

Fred jumped from the cab, took the ten-gallon drum from the trailer bar and drained the water off, then lifted the engine bonnet and stuffed rag around the diesel pump to keep it from freezing. He then put a sack over the engine, lowered the bonnet and locked the door. Fred had a word with the officer in charge, telling him we were off to find digs, and would be in touch. With that we made our way towards the town.

Most of the drivers had already booked in for the night. As we walked along the path, a woman shouted over to us: 'My sister will put you both up.' She handed Fred the address and told us it was near the Red Brick public house. We thanked her profusely.

Eventually we arrived at the house and knocked on the door.

'Come in and warm up, you both look half-frozen. It's not fit for a dog to be out in this weather!' the woman exclaimed. 'Dinner will be around five when my husband comes in.' Fred told her we were very grateful.

Later, as we sat having our evening meal, her husband told us that when the weather was really bad the police told drivers to park on that stretch of road coming into Warrington, and so he and his wife were used to putting up drivers for the night. When Fred told him how many lorries there were, he did not seem at all surprised.

It was lovely and warm sitting around the fire. Fred asked the gentleman of the house if we could spend another night with them if the police would not let us leave in the morning. His reply was that we could stay as long as we liked; his better half could do with the money anyway. They were such a very hospitable couple.

'What shall we do tonight, Fred? Stay in the warm or go for a pint?'

'Go for a drink. The Red Brick's not that far, so we'll suffer the cold,' he replied. 'We could be staying with these people for a week, Shay, so we have to give them space and privacy.'

Eventually we arrived at the pub. It was very warm and welcoming inside. A log fire was burning in the grate and the people were friendly. I offered to get the drinks in and sauntered up to the bar. The barmaid was talking to some customers; they were like bees round a honeypot and I understood why as soon as she walked up to serve me.

As I ordered two pints of best bitter, I noticed how slender her fingers were. She had a small waist, fairish hair, and she looked so feminine, very ladylike. As she gave me my change, I looked at her face; she was really beautiful with a lovely complexion. I would say her age was about thirty-something. Sitting down alongside Fred, I said jokingly, 'Fred, you're going to love me.'

'Why's that then?' he said.

'I've just found you a wife.' As we drank, I did not stop talking about the barmaid.

'Shay, I know you are only trying to wind me up. If she's as nice as you say and in her thirties, she's probably married with a couple of kids.' As soon as Fred had drunk his pint, I grabbed our glasses and walked towards the bar, noticing as I went that there was a piano in the corner.

'Same again?' the barmaid asked.

'Yes please,' I answered. 'By the way, is the piano working?'

'Yes. But it hasn't been played for a long time. Can you play?'

'No, but my friend does. He's really good. You can't see him from here.'

She realised I wasn't from those parts, judging by my accent. 'No, actually I'm from Kent,' I said. 'You have a charming accent yourself.'

'Thanks. I'm from Leyland,' she told me.

I walked over to the piano and put the beer on top. When Fred asked where his drink was and I mentioned the word 'piano', he shot off to try it out. As soon as his fingers touched the keyboard, he was away. He was a natural player. The whole pub fell silent and everybody looked up to observe the stranger sitting at the piano. Fred started playing modern songs and everybody sang along with him. I glanced towards the bar and could see that the barmaid was observing him as well.

The pub was packed out. An elderly couple had now joined the barmaid behind the bar. I could not get close enough to the counter to get served, so I shouted out, 'Two pints please, Lancs!' This drew her attention immediately and it was not long before she was serving me.

'Nobody has ever called me Lancs before.'

'Well, from now on I'll call you the Lancashire Lass.' I walked away smiling before she could answer.

I had just moved away from the bar when she called after me, beckoning me back. 'My dad said the beers are on the house for the night with our compliments,' she said, giving me my money back.

'Thank you. It's much appreciated by both of us.'

Fred looked up at me and in a quiet voice said, 'If I keep playing and we keep drinking like this, by the end of the evening we'll be pissed as arseholes.'

I walked up to the piano as Fred finished playing, 'There's an Old Mill by the Stream'.

'Fred, do we have to collect money tonight?'

'Of course not. You can't go around collecting money after drinking free beer all night. Use your loaf, for goodness sake, Shay.' I laughed and told him I was only joking.

It was a shame when they called, 'Time, gentlemen, please' because Fred was just getting into the swing of things. I spotted Lancs collecting glasses from a nearby table, and nudged Fred to draw his attention to her. At almost the same time, Fred and Lancs looked at each other; I could actually feel the chemistry between them. She blushed to the roots of her hair as she walked away.

'Isn't she lovely, Fred?' I said.

Fred agreed she certainly was, adding: 'Well, as we've had free beer all night, it won't hurt us to help her collect some glasses.'

I grinned at Fred: 'You are enthusiastic. It must be love.'

We had really enjoyed ourselves in the Red Brick; the time had passed far too quickly.

By now everybody had left the pub, and we stood talking to the elderly couple behind the bar. It turned out they were Lancs's Mum and Dad. After chatting for a while and thanking them once again for the beer, we bade them goodnight, making a special point of emphasising Lancs's name. Her parents and Fred looked at me so old fashioned, as if to say, 'Who's Lancs?' But nobody questioned me.

As we made our way back to the digs, the wind was blowing fiercely, and it was snowing so hard that you could not see properly. We kept our heads down and collars up. Before going indoors, we knocked the snow from our boots.

'So,' I said, once we were indoors, 'what do you think of Lancs, eh, Fred?'

'Not bad at all, Shay. She's a very classy looking lady.'

The bedroom was cosy and warm and with all that ale inside me, all I wanted to do was go to sleep. Just as I was dozing off, I thought to myself: if only I were a bit older, I would go for her myself. Suddenly, I heard a voice in the dark say, 'Shay; she's all right.'

The landlady knocked on our bedroom door the following morning at around nine o'clock, telling us our breakfast was nearly ready. Fred got up and looked out of the window. 'It's still bad out there, Shay. We'll have to telephone the police as soon as we've had our breakfast.' On opening the door of our room, we could smell the delicious aroma of food wafting up the staircase. The landlady did us proud with a hearty meal.

As we left, I thanked her for the breakfast, saying we'd probably see her that night. She called back, 'Cheerio, lads, have a good day.'

When we found a telephone box, Fred rang the police

who told us that it was not yet clear and to stay where we were. They would be in touch as soon as possible. 'Have we got to be out all day, Fred?' I asked wearily.

'Yes, it wouldn't be fair to them. We would only be in the way. Anyway, first things first, Shay. We'll check the lorry over.'

It must have been a good quarter of a mile walk to where we left the lorry. When we arrived at the park, we found that other drivers were already checking their vehicles. We stood chatting to them for a while, talking about where they were staying and how satisfied they were with their accommodation. Fred and I checked the lorry over; the ropes were frozen to the tarpaulin, and by now we were both beginning to tremble with the cold. 'Let's get going, Shay,' said Fred, and I wasn't going to argue with that.

'We're off now, lads,' Fred shouted to the drivers.

'Are you going back to your digs?' one of them bellowed back.

'No! We're off to the Red Brick pub,' Fred replied. The driver asked if he could join us and we set off.

As the three of us trekked to the pub the snow drifting onto the path made it more and more difficult for us to walk. By now the rest of the drivers were following on behind. I was so relieved to see the pub in the distance. We were frozen to the marrow.

The landlord was astonished to see us all troop in, especially so early in the morning, but it soon became apparent to him that the others were also stranded lorry drivers. I ordered two pints of best bitter and looked around the bar for Lancs who was nowhere to be seen. 'You couldn't give us a tune on the old piano?' asked the landlord. 'It will liven things up a bit. It's Fred, isn't it?' Fred was happy to oblige and was soon playing.

Then out of the corner of my eye I spotted Lancs. I guessed she'd arrived on the scene because she'd heard Fred playing. The pub soon livened up as all the drivers started to sing along with him, and when the locals arrived it was not long before they were singing as well. I heard the landlord say to Fred that in all the years he'd been a proprietor, he'd never seen it so busy at that time in the morning.

I thought to myself: what a great bloke. Turning to me, he said with a raucous laugh, 'Young man, I hope the snow is here to stay. It's certainly good for trade.' But it was Fred and I who had the last laugh because the beer was on the house for us.

Lancs had disappeared from view, but when she reappeared she had changed her clothes, and had a little make-up on. She looked absolutely ravishing, and I was spellbound. Most of the other drivers were ogling her too. It was plain they fancied her as well. But, unknown to Lancs, I was taking note that now and again she glanced over to where Fred was playing. I thought that was cute.

Since Fred was still playing, I decided to make myself useful by collecting some glasses. As I walked toward the bar, Lancs came over and thanked me. 'Your friend plays the piano well,' she smiled.

'I think Fred's the best player I've heard in a long time,' I agreed.

She asked if we worked together and I said we did, and that Fred was a terrific person and I enjoyed working with him immensely. 'He drives an eight-wheeler and trailer, and I'm his trailer boy. I love every minute of it.'

'His wife must miss him terribly when he's away from home,' said Lancs shyly.

'She was a career woman,' I replied. 'It didn't work out and they're divorced.'

'Were there any children from the marriage?'

'None. I think Fred would have loved kids.'

'Really! Unfortunately my husband died serving his country.'

In an effort to change this gloomy subject, I said, 'Oh I didn't tell you. The only difference between Fred and me is that I'm the good looking one, and Church of England, and he's ugly and Catholic.'

'A Catholic. I am too.'

Just then her father called out, 'Joyce, give me a hand.'

'I'd better go, Shay, he sounds a bit crotchety. He's been hard at it all morning.'

I went over to Fred and whispered in his ear, 'By the way, Lancs is a Catholic, and a widow, and her name is Joyce. But you didn't really want to know that, did you?'

He stopped playing and answered, 'I must admit, I wouldn't mind getting to know her better.'

At that moment, Joyce's father, Pete, rang the bell and shouted, 'Time, gentlemen, please.'

When the pub had finally emptied, Pete asked Fred and me to stay for lunch. We went into a room at the back where Madge, Pete's wife, told us to sit down and make ourselves at home.

When Lancs entered the room with a ploughman's lunch for each of us, she looked very shy and demure. Fred rose from his chair and winked as he took the plates from her and set them on the table.

'I must say, this looks very appetising,' Fred commented.

'Thank you,' she replied modestly.

Madge told us Joyce had prepared all the food. We could see by the way she looked at her mum that she wished the ground would open up. The poor girl was so embarrassed by this remark that she didn't know where to put her face.

After we'd finished our meal, Pete spoke about everything from steam trains to Lancaster bombers. We found him very interesting and friendly.

As we talked, Madge and Joyce started to clear the table. Fred offered to wash up if Joyce dried. Madge and Pete retired to the lounge for a rest before opening again in the evening.

I rang Banksy to let him know the police had not given us the all-clear yet, but that we would keep in touch. Then Pete invited me to sit with them until Fred and Joyce had finished the dishes. I was happy to accept but I still quite envied Fred.

Pete showed me all his books. Mostly they were about things that happened during the First World War. It wasn't really my subject, so I was hoping Fred would rescue me soon. Madge must have been reading my thoughts because she said, 'I don't know about you two, but I could do with a cup of tea. I'll go and put the kettle on.'

Soon Joyce came in, followed by Madge and Fred – much to my relief. When we had drunk our tea, Fred thanked them for their hospitality and said we had better be making a move, but that we might see them that night, depending on how the roads were.

The weather was still appalling. As we neared the

lorry to check it over, the police were there talking to another driver. They told us that a snowplough had cleared the dart board on the A6 and we could be on our way within half an hour. Fred reported to the patrol police, then we made our way back to the digs to pack our cases.

At the pub, Pete and Madge were half-expecting us at the pub that night. 'Well, never mind,' said Pete. 'Drop in if ever you're this way again.'

'You try keeping us away,' Fred replied. Just then Joyce came out. Fred told her he would be in touch. I got the impression Fred wanted to talk to her alone, but it was difficult as time was getting on. They wished us well and all three waved as we walked on up the road with the wind howling around our ears.

When we approached our vehicles, we could see some of the drivers giving the council men a hand to shovel the snow away from the lorries. The ten-gallon drum of water which I had put inside the cab when Fred had parked up was almost frozen. Dragging it out, I took it across to where the council men had a coke fire burning away merrily in an old can with holes in it which rested on top of some bricks. It made Fred and I feel warmer just looking at it.

Most of the drivers had put their drums around the fire to thaw out. One of them remarked, 'I don't know what we would've done if they hadn't lit that fire.'

In the meantime Fred checked the lorry over. As soon as the water had thawed, Fred quickly poured it into the radiator, jumped up inside the cab and hit the starter button. That old Gardner engine started like a dream.

Fred left the lorry running while we walked over to some of the other drivers to make sure their vehicles were okay. For those that didn't start, Fred put a naked flame down the breather tube while I used the starting handle. Some of the older lorries were bastards. They just did not want to start up.

At last all the engines were roaring, and the drivers were ready to move. The police said the road was almost clear, so we followed one another onto the A6, which was the same width as the plough.

'This is ridiculous,' Fred commented. 'I think we've dropped a bollock leaving Warrington, Shay. I can't see us getting very far in this bloody lot.' But there was no turning back now. The council had done a good job of salting the road, but on either side there must have been a good three feet of snow.

The old Gardner was barking beautifully. However, at nine miles an hour, it was going to be a long haul to Lancaster. I was praying the roads would clear soon. If the lorry got stuck we would be in deep trouble, to put it mildly. Sleeping in the cab would be no joke.

All of a sudden Fred had to brake, and we slithered to a halt. We sat there for at least a quarter of an hour. 'I wonder what's up now?' Fred remarked.

'How the hell should I know?' I replied. By this time I was feeling very uptight and edgy.

Fred jumped down from the cab to find out what was wrong. After five minutes he was back to tell me that about five lorries in front of us, the radiator of an Albion had blown up. The lorry in front of the Albion, a five-ton Bedford which was fully loaded, was going to tow him. 'If that Bedford pulls twelve tons at this speed, Shay, you can bet your bottom dollar it will blow up within five miles. Which means we'll have two lorries in front of us broken down.'

‘Great,’ I replied, feeling even more irritable.

‘It’s being so bloody cheerful that keeps you going, isn’t it, Shay?’ As Fred pulled away again, I could see out of the corner of my eye that he was finding me very amusing. I just sat there feeling sorry for myself.

By now the snow was beginning to melt. The road was very slushy and wet, but it wasn’t so treacherous. Fred still had to keep his wits about him all the time. The road was gradually widening as we cruised along, which made it easier. Fred threw me a glance, smiled and did one of his famous fancy gear changes, then pulled out and overtook the lorry in front. Our speed was nearing twenty miles an hour.

As we passed I noticed that the Bedford had stopped on the side of the road, with steam belching from its radiator. Another driver had pulled in behind him to give him a hand. By now I was feeling better because we were well on our way to Lancaster. All I could see in the mirror was the spray bouncing off our wheels, making the windscreen very obscure.

It had started to warm up in the cab; my toes were gradually coming back to life. Fred put the headlamps on as it was getting dark. By the time we reached Lancaster it was nine o’clock. It had taken about seven hours, which wasn’t bad going. We managed to find digs all right. The landlady even cooked us a meal, which went down a treat as we were both starving and our throats were parched. Straight after our meal we climbed into bed, completely and utterly exhausted. It had been a long, eventful day.

In the morning, after a hearty breakfast we left the digs at around eight. Fred headed straight for the builders' merchants near the railway station where we unloaded the lorry and trailer.

'I'll ring for a return load,' said Fred, 'but I'm not taking anything that has to go across to the other side of the country. I want one that keeps us here. The snake passes are a nightmare in this weather.'

'It's not so bad here, but it could be a darn sight worse, as you say, on the other side,' I replied.

'That's settled then, Shay. We'll keep well away from Yorkshire.'

When Fred rang the clearing house, the manager there said he had a lot of work owing to a few drivers still being stuck in snowdrifts. Fred and I loaded eighteen tons of cotton bales for delivery to London. As Fred pulled out of Lancaster, I noted the time: half past one. We had done well to load and tip all in one morning.

Fred wasn't sure where we were stopping that night. 'I know what I'd really like to do,' I volunteered. 'Punch on to Warrington. That way we would be able to see Lancs and you could play the old Joanna.'

Fred seemed to agree with this, because he clean forgot about the icy weather conditions and put his foot down so hard on the accelerator that I hung on for dear life as we slid all over the road. When the lorry went to the right, the trailer pushed it more, making it difficult for Fred to pull out of the skid. He was laughing, but I was frightened half to death. I must admit, it broke the monotony a bit. Eventually Fred managed to control the lorry.

Every few yards there was an abandoned car in the snow; I had never seen so many. Some of the car drivers stayed with their cars up to the last minute, hoping they would be helped, but many had given up hope before dawn and just abandoned them. The only people who didn't seem to worry were reps with company cars.

When we were approaching Broughton we spotted a slight incline. In the distance we could see drivers standing with their lorries, stuck fast, beating their hands across their chests trying desperately to keep warm. As we drew closer, some of the drivers shouted over to us, 'Keep going' and 'Don't stop!' Fred kept his foot down. About halfway up the hill, Fred began to ease up, as the tyres had started to lose their grip in the snow, and the lorry started to shudder.

'Come on, girl. Come on, girl, you can make it,' I shouted, willing the lorry to make it to the top. A few feet from the summit we very nearly came to a standstill. The eight-wheeler started to slide broadside. Fred only just managed to straighten her up. We both whooped with delight. I hadn't thought we would make it with a trailer on the back.

'It was the double drives that got us up there,' he told me.

I had noticed that Fred had wound the ratchet hand-brake up. Was that in case it ran backwards? I asked. 'No, Shay. It's to stop the back wheels spinning. The slower the wheels go round the better the grip.'

'I've certainly learnt a lot since working with you, Fred.'

'Just stick around, kid, and you'll learn even more.'

I felt sorry for the lads who were stuck on the hill, but they realised we couldn't do anything to help, especially with a dangler on the back. That's why they told us to keep going. Although Fred drove slowly, we made steady progress. Eventually we arrived in Warrington, stopping at the same digs as before. No sooner had we eaten than we washed, shaved, and in no time at all were heading off towards the Red Brick pub.

As we walked in, some of the locals acknowledged us, remembering Fred's piano playing. Pete was delighted to see us. Fred ordered two pints of bitter. I looked around for Lancs but she was nowhere to be seen. To requests of 'Give us a tune', Fred walked over to the piano, sat down and started to play.

Suddenly, out of the corner of my eye, I caught sight of Lancs. Then she vanished. After about a quarter of an hour she reappeared, looking as bright as a new penny and so exquisite. She walked over to the piano, said hello and how lovely it was to see us again, but all the time she was talking to me, I knew she only had eyes for Fred.

As she stood by the piano, Fred asked: 'What's your favourite tune, Lancs?'

'"As Time Goes By",' she said, 'from the film *Casablanca*.' Fred started to play it as she walked back towards the bar. She turned and looked at him, smiling. Fred winked at her. He seemed to be putting more effort into that song than into all the others that I'd known him play. Lancs was definitely having some influence over him.

After he had finished playing, Fred looked towards the bar, and caught sight of her lips miming 'thank you' to him. The people in the pub that night seemed to be enjoying themselves immensely. It was turning out to be a wonderful evening; Fred really made that old Joanna sing. I was glad we had stopped off at Warrington – and Fred was too.

Soon Lancs walked daintily over towards us – she was so feminine, it was untrue – and she was holding a plate piled high with sandwiches. 'These are for you and Fred,' she whispered.

'Dear, oh dear, Lancs. There's enough here to feed an army,' I replied. But they went down a treat with the local bitter.

Later, as I thanked Lancs at the bar, I heard one of the customers say, 'Oh no, here comes Rent-a-Mouth.'

Lancs whispered to me, 'I can't stand that man.'

A tall man with hair swept back had just walked in and was being served a beer by Pete. The newcomer had a pencil-thin moustache and an air about him which said, 'Look, everybody, I'm here.' I immediately saw why Lancs disliked him so much.

From talking to Lancs, I found out that moustache man worked for Avery, setting the weighbridges up, and was away more than he was at home. I could feel myself getting angry because as Lancs and I were talking he constantly butted in on our conversation. I could feel the weight of his body trying to push me out of the way, so he could be nearer to Lancs. My temper was rising, but I controlled it.

Lancs walked to the other end of the bar to serve a customer because I think she sensed what was going to happen. She explained to Madge what was going on. After a while the man sauntered up to where Lancs was, so she deliberately walked back down towards me and left him standing up the other end of the bar. 'Does he fancy you, Lancs?' I teased her.

'He fancies anybody in a skirt. He's probably got a wife and kids at home. The other customers don't mix with him because he knows everything but nothing, if you get my meaning, Shay. And he's a bully to boot. The trouble is, he seems to win the arguments.

He'll get his comeuppance one day, you mark my words.'

By now some of the locals had joined us. One of them remarked how much he enjoyed Fred's playing; another said how it brightened the atmosphere of the pub. Then, at that precise moment, Lancs's head turned towards the piano. I could hear Fred playing 'As Time Goes By' for the second time that night. All of a sudden, Rent-a-Mouth made his presence felt again. Some of the regulars walked back to their tables, not wanting to know him.

He walked over to Lancs, introducing himself to her as Ian. I overheard him making suggestive remarks to her. I was becoming angrier and angrier by the minute; he was being offensive to her now. I couldn't stand it any longer, so told him point-blank to go forth and multiply. How I stopped myself from swearing, I don't know. All he could think of was to call me a young whippersnapper. I didn't know at the time but Fred had left the piano and was watching what was going on.

Suddenly, without warning, Ian took a swing at me. I ducked, fell backwards and hit my head on the table, which inevitably sent the glasses flying. His punch didn't actually touch me. I got up, rubbing my head. As Ian came towards me again, arms flailing, he was stopped in his tracks and yanked back by his collar. As he turned to see what was going on, Fred punched him straight in the stomach. Ian fell to his knees, gasping, his false teeth sticking out of his mouth. The power behind that punch must have been equivalent to a 56-pound hammer.

Fred put his hand on the man's shoulder. 'If you ever, I repeat, if you ever touch my friend again, make sure you order a six-foot box first. Now piss off.' As the bully stood up, Fred grabbed him by the seat of his trousers and collar, shouting out to me to open the door. He then frogmarched Rent-a-Mouth and threw him out.

When Fred had made sure I wasn't hurt, he said sadly: 'Well, Shay, I expect we've blotted our copybook now. They won't want us in here any more. I think we'd better creep off quietly.'

'We can't do that Fred! That feller deserved it. He can't go around talking to women like that. Come on, I'll buy you a beer for a change. Anyway, I know one thing: Lancs and the customers don't like him, either.' This seemed to cheer him up a bit.

As Fred and I walked up to the bar, Pete and everybody in the pub started clapping. One of them shouted out, 'Well done. He asked for that.'

Fred sat down at the piano and played 'As Time Goes By' again, and from that moment on Lancs and I started a great friendship that has lasted all our lives.

I watched covertly as Lancs walked over to the piano with a pint of beer for Fred. They talked for a long time. Even Pete and Madge noticed the way their daughter was looking at Fred. Pete winked at me approvingly. I could see from their expressions that they had taken a liking to my driver.

Pete rang the bell for last orders. As the locals began to leave, some called out to us, saying they hoped to see us again soon.

Pete locked up while Madge made us all coffee. Afterwards, Joyce saw us off the premises. As we started to walk away, I said in a hushed voice to Fred, 'Aren't you going to give the lady a goodnight kiss?' Much to my surprise he walked back and kissed her on the side of her cheek.

As we trudged along in the snow, Fred turned to me. 'Keep your eyes open for the fellow I threw out of the pub. He may be loitering around somewhere and jump us in the dark.' As soon as he said that I was immediately on the alert. However, I needn't have worried; the man was nowhere to be seen.

The next day it was back in the saddle punching south. The weather was still unsettled. As Fred drove back down the A50 it started to snow yet again. The windscreen wiper that Deafy had fitted on my side helped a lot. When the snow built up on the blades, every so often Fred would stop when it was safe to, and I'd jump out and tap the snow off.

We didn't stop at transport cafés in case the inevitable happened. Fred just kept driving and hoping for the best. The towns on the way through looked very picturesque, but driving was still very dangerous.

Suddenly the engine started to fade. Fred told me the diesel was waxing.

'What does that mean?' I asked.

'It means the diesel is freezing.' Fred pulled over while I muffed the radiator up with a sack. Then he lifted my side of the bonnet and stuffed rag around the fuel pump, keeping the engine running all the time. Within a few miles the diesel stopped waxing and she was pulling much better. Fred pulled in again and I dropped the muff half-way down the radiator. When it was all done, we pulled off again. I was freezing to death in the cab.

As we drove on I probed Fred about Lancs. Did he have a soft spot for her?

'I've never met anyone quite like her, Shay. She's gorgeous. But only time will tell,' he smiled.

It was proving to be a long day. The weather conditions hadn't improved at all. We eventually stopped at the Watling Street café in Markyate, one of the main transport cafés. I drained the water from the radiator, while Fred derved up and checked the engine oil.

We walked through the door of the café carrying our suitcases. It wasn't one of the best places to stay, but the beds were clean, and you could always get an early morning call. Fred and I decided to stay in that night as we felt worn out from travelling and it was still bitterly cold outside.

The following morning at four, I filled the radiator, muffed it half-way, and scraped the ice off the windscreen. Fred started the engine, and left it running until it warmed up a little. We were going to make our way to the Surrey Docks in the East End of London. It was really touch-and-go driving around the London colony, but in no time at all we were going through Barnet. The roads in London were remarkably good as the buildings had helped shelter them. As Fred drove into the docks, he told me that sometimes drivers were kept hanging about for days as the dockers were lazy bastards, and didn't like work at all.

We were among the first to arrive. We reported to the Checker's Hut, and to our amazement one of the men came round with the 'grab crane' to off-load our bales of cotton. We couldn't believe our luck. Pulling out of the docks, Fred looked over and said, 'I take back all that about the people in the docks. I've never been in and out so quickly.' As he drove out of London we headed towards the A2. It was nine in the morning.

I was puzzled. 'Fred, how is it that we've delivered cotton to the docks? I thought it was imported in, not exported out.'

'Shay, we live in a crazy world, but as long as we get a good rate of pay for the job, Banksy's happy and, much more importantly, it keeps us employed.' Fred wheeled the lorry and trailer into the Merry Chest café, which was packed with other lorries. What a laugh we had in there with some of the other truckers. We must have been in there for about three hours.

All the North Kent drivers were rallying together and forming their own branch in the union. The drivers stuck together on all issues. I thought Charlie Earl, one of the local drivers, spoke a lot of sense. Over the last decade they had had enough of some of the working practices, such as driving old bangers. Their wages were low, on top of which the money they were paid for nights out only just covered their stay in digs.

Fred telephoned Banksy who told us to come back to the yard, praising us up because he had not expected us back so soon. As soon as we arrived I noticed that no more had been done to Deafy's new garage, due to the bad weather. All the bricks lay covered with about an inch of snow.

While we waited to see if the governor had any work for us, Deafy took the brakes up on the lorry and trailer. The governor eventually came out of his office, saying we were to go to the London Brick company as early as possible the following morning, and load.

The question was: how would I get back to the yard at that hour in the morning? Fred suggested that Banksy paid me for a night out – and himself as well, while he was at it. Banksy didn't like it much but he agreed.

Soon Fred had parked up about a quarter of a mile from Banksy's yard in Lowfield Street car park, so we had a night out with pay in Fred's digs.

The following morning we set off. As it was so early it didn't take long to get through London. Fred was soon driving along the A5. On the other side of Dunstable, he turned right and drove along the A5120 through Toddington and Ampthill to a place called Stewartby. I remember going under a very low arch, and thinking the top of the cab would hit it. We eventually came to a large estate where most of the workers from the London Brick company lived.

We reported to the office. The fellow there told us that most of his lorries were stuck in the snow somewhere. I asked cheekily if there were any deliveries for Warrington. The old boy gave a belly laugh, saying, 'Don't you people ever want to go home?' He told us they were screaming out for bricks in Brighton.

The Polish workers loaded us quickly, being very adept at their job. Fred and I protected our eyes from the brick dust with our hands; the wind didn't help, it was blowing a gale that morning. The Polish guys had their eye-shields on. Then we reported back to the gate-house. The checker, who had been doing the job all his life, just looked at the lorry and knew immediately how many bricks we had on.

We planned to stop in Dartford later, but book off somewhere else on a dodgy. I was learning the tricks of the trade fast from Fred. 'Oh, by the way, Shay, don't forget to bring a clean set of underpants with you. It's beginning to notice.'

'Saucy sod,' I laughed.

When I eventually arrived home, Mum was standing

in the kitchen rolling out some pastry. 'Hello, son, it's nice to see you.'

'You, too, Mum,' I replied. 'What's for dinner? I'm starving.' It was my favourite: meat pie. I loved my mum dearly but she was constantly nagging me about working away from home, saying, 'Dad's a lorry driver but he's home every night.' It was hard to convince her that this was going to be my way of life, and I enjoyed it. Although Dad liked being at home with her every night I didn't have to be the same.

While Mum was serving dinner, Dad walked through the door. He was pleased to see me safe and sound, like always. As we talked, Dad said he would like to meet Fred Ruddock, as I was always talking about him.

After dinner I handed Mum some house-keeping, and her face lit up. 'Thank you, son, I could do with that.'

Poor old Dad nearly choked when he saw it. 'Blimey, lad, that's over two weeks wages for me.'

That was typical of the difference between a transport man and an ordinary driver. I explained that Fred and I did a bit of fiddling now and again, and being on the road all the time made it easier.

'But you can't do that, Dad, because as sure as God made little apples, you'd get caught.' I turned and gave Mum a kiss, thanked her for a smashing dinner and headed for bed.

I met Fred early the following morning with my suitcase packed full of clean gear. Working with Fred, I didn't know where I was going to end up next – wherever Fred hung his hat was his home.

It was still bitterly cold but the council had gritted the roads well, which made driving conditions a lot better. Fred drove through Tunbridge Wells on course for Kemp Town, Brighton, where there was an enormous house-building site. Owing to the weather there was only a skeleton staff.

When we started to unload, one of the building site workers came over to us and asked if we could deliver some bricks for him privately, adding that he would make it worth our while. He lived about a mile away. Fred told him that as long as we had his signature on our ticket, there wouldn't be a problem. 'How many do you want delivered?' Fred asked.

'What you have on your trailer will do me fine,' the builder replied.

'Right, we'll unload the trailer from the front end, then you can tell us when you have the bricks you need.' Eventually the builder told us that the amount we had left on the lorry would do him – eight tons.

We unhooked the trailer, making the steering much lighter for Fred to manoeuvre. The reason Fred had worked it so that the remaining load was on the back end of the lorry was just in case we hit a patch of ice.

When we reached the housing estate, I saw a few net curtains twitch as Fred drove up the road, because the old Gardner engine threw out a lot of thrust, vibrating the windows as we went. Fred pulled up behind a small red car. As we jumped down from the cab, Fred immediately started to unload. I watched in amazement, never having seen anyone carry so many bricks all in one go. His arms were stretched to the limit. 'Come on then, Shay, don't just stand there gawking. Give us a hand.'

The quantity of bricks I could carry was nowhere near the amount Fred was managing. We stacked them on the path. When we'd finished, the builder offered Fred a fiver.

'Make it six. That's three pounds each,' Fred replied.

The builder agreed grudgingly; I got the impression he did not want to part with too much money. He signed our ticket and thanked us.

'Three pounds each, not bad for a day's work, is it, Shay?' I agreed, wishing we could do it every day. As we jumped back up into the cab I remarked that we'd just earned ourselves more than we got in wages. Ever the pragmatist, Fred replied, 'Shay if we earned more, we wouldn't have to fiddle to get by.' He was right, of course.

When we got back to the building site and had hooked the trailer back onto the lorry, Fred telephoned Banksy. As he climbed back into the cab, he told me we would be having another dodgy night out, as we had to load out of Reed's Paper company, which was in Maidstone, and take reels up to Newcastle-upon-Tyne. 'It means we'll be working a full seven days again, running bent as arseholes,' Fred laughed. 'But if we have to go away, we might as well work the maximum hours and earn well doing it.'

With most firms, if you were away you usually parked up at weekends. But Banksy liked his drivers working the full seven days. 'That way the lorry keeps earning him money,' Fred said. 'He's a crafty old bastard.'

Fred told me that if I needed a break, he'd park up somewhere, and we'd have a couple of days off in any town (within reason) that took my fancy. I couldn't complain about all the work: since working with Fred, I'd earned a lot of money. 'I didn't tell you Shay,' he said, 'but last week I booked seventy-seven hours, and seven nights out – that's for both of us, mind.'

'Banksy'll do his nut, won't he?' I said.

'As I've said many times, Shay, I'm the captain of this ship, and if Banksy doesn't like it, he can do the other thing.' Of course, this meant, 'He can sack me.'

'I'll tell you this, Fred, I won't work with anybody else. If you leave, so will I.'

'That's what I like to hear. If that's not comradeship, I don't know what is,' he replied. He then made me promise not to mention what we did to anyone, as it would cause a lot of problems. Meanwhile, I'd already made up my mind to open a post office account and put all the money I'd earned into it. When the time came to leave, I'd have something to show for it.

Suddenly I asked Fred if he would like to meet my mum and dad, and much to my astonishment he agreed. He parked the lorry and trailer in Bath Street, near the Gravesend Ferry.

Mum was in the garden when we got there. I could see from her face that she was flabbergasted to see me, especially after packing my case the previous evening. I introduced Fred to Mum, and bless her, she got a bit flustered, trying to remove her apron, and shaking Fred's hand at the same time, saying it was nice to meet him at long last.

She asked Fred if he would like to stay for dinner, and he said he would love to. I was so relieved when she asked him, and even more so when she said Fred could sleep in the spare room. My old mum was great; she never once let me down. Whoever I brought home was made to feel very welcome.

I looked across at Fred and said, 'Not bad, eh! Dodgy night, and free digs.'

When Dad arrived home, we all had dinner and chatted. I could see immediately that my folks had taken a liking to Fred. It was well past midnight when we finally went to bed.

The following morning, Dad was up and gone – the early bird catching the worm. Fred and I had a lie in, as there was no immediate rush. With no bosses watching over us, we worked hours more or less to suit ourselves.

It was eight-thirty when we finally got up. Mum set us up for the day with a good cooked breakfast, fussing over us like an old mother hen. As we left at nine-thirty, I saw Fred kiss her cheek and press something into her hand.

Chapter 5

At Last

IT took us about half an hour to get from Mum's to Reed's in Maidstone. When we arrived at the mill, Fred saw one of the workers and cadged a few scotches and two back boards from him for the lorry and trailer. While we chatted to some of the other men, the driver of the overhead crane loaded us. I roped and sheeted, while Fred filled out a log sheet in the cab, booking us on at six that morning, and booking a break for half an hour. Then he said, 'Newcastle, here we come,' and we both knocked off the ratchet handbrakes.

It was around midday when we left Maidstone, and I was worried we were a bit late. 'That's no sweat, Shay,' Fred told me. 'We've got a couple of days to make up for the time we've lost.'

Fred pulled onto the A20 and headed towards London. As we drove through the village and towns, it was a hard old slog as it was single lane most of the way. The whole journey took six hours to where Fred had booked our digs in Grantham. We didn't even stop for a drink. The A1 was slushy where the snow had begun to thaw, so the roads weren't too bad.

Banksy was a funny old bugger. His lorries were in tiptop condition, a hundred per cent. They looked second to none, with beautiful sign writing and un-believable coach-work. However, his drivers always ran bent, much to his delight and encouragement, with pay. He was an excellent governor in his own right, but very tight. They said that during the war he had driven a lorry for twenty-four hours non-stop, which would have been a feat in itself. Whether it was true or not, nobody knew.

As we drove along the Al I looked at the lorries parked alongside the road. The condition of some of them was absolutely dreadful. The one that really caught my eye was a Reo Speed Wagon Articulator with a single-axle trailer. The bonnet was very long. It looked gigantic. It was loaded with slabs that were almost touching the road, because the trailer had completely broken in the middle.

'The only trouble with them, Shay,' Fred told me, 'is they're fine when they're empty, but when they're loaded they haven't got the strength to pull your foreskin back. That driver is what's known in the trade as in the shit.'

There was a lot of crap being driven in those days, but there were also some great trucks. In the heavy class there were Atkinsons, ERFs, Scammells, Leylands, AECs and Albions. They were supreme, especially if a Gardner engine was under the bonnet.

Eventually we arrived at Tony's in Grantham. As Fred pulled in, I remembered the laugh we had had with some of the Scottish lads when we were last there. As we sat chatting and eating our evening meal, who should walk over but Big Ian. 'How's that trailer boy of yours, Fred? Is he still the same as you told me last time? Bloody hopeless.'

I stood up, raised my hand and said, 'If you want aggro, mister, step outside.'

The other drivers started laughing. Big Ian patted me on the shoulder, which propelled me three paces forward. 'Come on laddie,' he said. 'Let's go down the pub for a drop of the hard stuff.' The rest of the lads decided to join us.

The landlord was a happy man to see us all walk in. He told Fred his takings had doubled the last time. As we approached the bar, Big Ian suggested having a kitty. All the drivers agreed, so each of us put in half a dollar – two and sixpence was a lot of money in those days. The popular drink was brown and mild, and also bitter.

One of the locals was playing the piano, hitting the wrong keys at times. The poor old soul was a trifle flat, just like my mother's frying pan, but he tried. The next thing I knew, Fred was walking over to the piano and tactfully offering to let the man have a break. As soon as Fred sat down and started playing, the mood in the pub became more lively and some of the customers sang along with him.

Once again we all had a good night. When the landlord rang the bell and shouted 'time', Fred started to play 'There's an Old Mill by the Stream'. That was my cue for the night; I did my usual stint of going round with my pint pot, saying 'Don't forget the pianist.' That night I got my two and six back, plus.

As we all ambled up the road, everyone said how cold it was. There was still some snow lying about. Driving trucks in the winter was a nightmare. Quite a few drivers caught pneumonia. One driver was found in his cab the other side of Stratford-upon-Avon, dead from hypothermia. Big Ian scoffed, 'Some of you don't know what cold means. You want to go to Scotland, where I live.'

'That's not quite true,' Fred responded. 'Most Scotsmen are down in the south, driving. It's all we southerners who are up in Scotland having to put up with the bad weather.'

Back at the digs, I switched on the wireless. The weatherman was saying it would be mainly dry with some rain or snow showers later. I thought to myself: that covers it all. Fred and I didn't bother with a nightcap. As it was late we just went straight to our beds.

The next day Fred had a bit of a burn-up with Big Ian as we punched along the Al. Big Ian grinned as he overtook us in his eight-wheel Thornycroft. We had the dangler on the back, which in theory made us a lot slower – but all the same Fred just managed to keep up.

The condition of the road was still far from good. As we went further north, there was a flurry of snow but nothing compared with what it had been. Eventually, after a long haul, we pulled in at Scotch Corner, hoping they'd got a fire burning in the café. As we walked in, one of the drivers shouted over to us, 'What're the roads like further down?'

'Not too bad,' Fred answered. 'How about the roads up north?'

'You'll be all right as far as Newcastle, but the other side heading towards the borders of Scotland is a bit gruesome, and many of the roads are closed. Five minutes after the ploughs clear the road, it's just as bad.'

Later, we wished Big Ian good luck and shook hands with him as he was going to Penrith. Neither of us envied him the journey. He told Fred to be extra vigilant, especially with a trailer on the back. Fred and I jumped up inside the cab and were soon making tracks towards Newcastle.

On the approach to Ferryhill, about twenty miles from Durham, it began to snow heavily again, making driving difficult once more. We had just started to go up a hill when the wheels of an ex-army, square-nosed petrol Bedford, which had a three-ton Scammell trailer on the back, started spinning. The front of the lorry and unit jack-knifed with the unit slewed half-way across the road so that it was impossible for the trucker to drive up or down.

When the trucker looked out of the nearside window, all he saw was the trailer and the traffic behind him, and when he looked out of his side window, all he could see was the brow of the hill in front. He was in a right old predicament.

I could sense Fred's impatience and annoyance as he pulled up behind. He was completely pissed off. I sat there thinking to myself, 'Shay don't say a word, otherwise he will be telling you to get on a train and fuck off back to Kent.'

Behind us there was mayhem as the vehicles stacked up. Some of the other drivers were becoming agitated.

We wound the hand-brakes tight and climbed down from the Atkinson. The Bedford and trailer were still in army colours. It had a small headboard with Rosswell's Transport written on the top. Fred approached the driver: 'Looks as though you could do with some help.'

'You could say that,' the driver replied.

'There's only one way to get out of this mess. Have you got a thick rope?' As the driver went to fetch it, I could see Fred's brain ticking over as he looked at the vehicle. Fred then put the rope on the offside spring hangar, on the back end of the trailer. 'Shay, don't let anyone squeeze past, otherwise it will make matters worse.'

He jumped up into our cab and pulled across to the other side of the road, turned our lorry round, then drove back to the right side of the road, straight onto the other trailer. Fred was now half-way across the road. He hooked the other end of the rope onto the tow pin on the front of the Atkinson. Then he asked me to knock the hand-brake off.

With that he started the lorry up, put her in reverse and backed down gently, pulling the back end of the other trailer right round. It slid easily on the snow. Fred then jumped back down from the cab and unhooked. He shouted to the driver, telling him not to drive up the hill, but to reverse back down and get the lorry as straight as possible.

Fred pulled in front of him. I hooked the rope onto the Bedford and jumped back inside the cab. Slowly, and with some anxiety, both lorries reached the top. I undid the rope, jumped back inside the cab and, as we moved off, the driver gave Fred the thumbs-up sign. To our relief it had all gone according to plan.

As we drove along I asked Fred if he'd learned that trick at Pickfords heavy haulage. 'No, Shay, it's just common-sense,' he replied.

Not long after leaving Ferryhill, Fred pulled in at Durham, where we would be parking up for the night. Fred had been there many times when he worked for Pickfords, and had got to know all the short cuts. I loved listening to him and asked him to tell me more. 'All right,' he said. 'Pickfords was a really well-organised company.'

He told me how he'd once left Newcastle with over

seventy tons on board. That was a lot of tonnage, and he was delivering it to the Ministry of Defence in Dover.

On another occasion, he'd carried 170 tons 116 feet long. On the front end were two Scammell lorries connected together with a tow-bar, both pulling the trailer, while another Scammell pushed at the rear. All three had Gardner engines. These lorries were especially adapted and built solely for Pickfords. The trailer had eighty-two wheels, and was specially designed to carry the weight.

Pickfords had installed full communication on all their lorries, so the drivers could talk to each other. On the front and rear end of the trailer were two small cabins which were both manned at all times. They had specially adapted front and rear low steering, which helped manoeuvre the trailer better.

Pickfords also had a five-ton Bedford which had a petrol engine. It carried all the spare parts, such as railway sleepers, chains, tow ropes and toggles. At the end of each day, when Fred and the other drivers left the lorries, they used the Bedford as a taxi back to the digs.

When they were taking their exceptional loads through towns they demolished traffic lights, beacons and metal railings on the side of pavements. You name it, they knocked it down – accidentally, of course. 'It must have cost the firm a fortune,' I said.

'No, Shay, it was charged to the customer's insurance,' said Fred. 'It was all done in the presence of the police. Fortunately, one of the crew was an electrician, so when we damaged electrical appliances, he did a temporary repair until the police contacted the right authorities. Our worst nightmare was hump-backed bridges, which involved a lot of bloody hard work – and I mean hard work. The trailer was so low on the road that when the first two lorries went over the brow, the trailer locked into the tarmac.'

Fred was now parked up. 'Well, Shay, I think I've bored you enough for one day. Drain the water off, and I'll take you to the digs where we all used to stay. We haven't booked in, but we'll take a chance. I know the landlady, so we should be all right.'

When we arrived at the digs, I could see from the landlady's face that she was very pleased to see Fred. Her name was Glad and her accent was one that I'd never heard before. Just listening to her speak fascinated me and she loved reminiscing.

She told me about some of the Pickfords lads who had stayed there during the war. 'When they arrived smothered from head to foot in grease and oil, they looked tired and stressed out,' she told us. 'When they'd finished their meals, they would always thank me and say how enjoyable it was. More often than not they would give me a hug and tease me in fun.

'Some landladies refused to take them in,' Glad continued, 'because of the grease and oil. But when they'd washed, shaved and changed into their casuals, you would never have recognised them. Often they would give me a bit extra cash. I loved to see them leaving in the morning, well-fed, looking nice and clean. Pickfords' lads were the best.'

After chatting for a while she showed us to our room, which was very clean with crisp, white sheets. 'While you two are getting yourselves cleaned up, I'll get the dinner on,' Glad said, and in no time at all we were tucking into a hearty meal. We then headed for the pub, telling Glad that if she was at a loose end she was welcome to join us. 'I might just take you up on that, boys.'

The pub was friendly. After a few pints, Fred told me that I was downing it much too fast and to slow down. He was right, as always. I took a lot of notice of Fred, whereas I probably would have had a go at anyone else saying that to me.

'If anybody offers you a Newcastle Brown ale, Shay, turn it down,' he warned me. 'If you drink too many of them, you'll not be able to control yourself!'

Suddenly the door opened and in walked Glad. She knew everybody. Fred was the first to offer her a drink. 'Oh, I'll have a gin and tonic, thanks,' she said. She took Fred by the arm and led him to the piano. 'Come on, let's have a sing song like we did in the old days.'

With a few gins under her belt, Glad soon started dancing to 'Knees up, Mother Brown', showing off her panties and suspenders. The locals told me it was her favourite party piece. She really was a great lady. Glad was soon joined by a few other women showing off their bloomers; I've never seen so many passion killers in my life, they were all colours! The Geordies sure knew how to have fun.

Fred played all evening, only stopping when he was lifting his right hand. Despite what he'd said, I downed quite a few pints, becoming a proper little piss-artist. Then suddenly I heard the old tune, so I grabbed my pint pot and staggered round the bar, slurring, 'Don't forget the pianist.' Drink gave me a lot of Dutch courage. I started actually telling the customers to part with their money, something I would never have done if I'd been sober. Even the landlord chucked in a dollar.

Poor old Glad, meanwhile, was so drunk she could barely stand on her feet. Fred and I took her by the arms and helped her keep upright. As we opened the door we were greeted by a blizzard, which surprised us, because they hadn't forecast any more snow. Well, they got it right most of the time, I suppose.

Back at the digs, Fred made Glad a strong coffee, then removed her shoes, helped her up to bed and threw a blanket over her. When he came back downstairs, he said, 'Make some tea, Shay, while I count the money. By the way, how many pints did you have?'

'About three,' I answered gingerly.

'You lying little toe rag! I counted at least seven, and since you've lied to me, I'm not sharing any of the kitty with you. Men who work with me must be honest and honourable. What do you have to say about that?'

'There's only one thing I can say, Fred. You're nothing but a miserable old Catholic bastard, and you can stick the kitty right up your arse.'

'Touchy, aren't we?' You don't honestly think I'd do a thing like that do you, Shay?'

'Not really,' I answered with a grin. He then gave me half the beer money. 'Seriously, though, Shay, watch your drinking. You're still a young man, and when you're on the road on your own, you won't be able to afford to drink like that. If you did, you'd be risking your job, and the money from your nights out wouldn't cover your expenses away from home.' Even though I was drunk, I never forgot Fred's words, and I always watched my drinking after that.

The following morning, Fred woke me from a deep sleep, saying he had cooked breakfast. In no time at all I was sitting down at the table tucking in to a delicious feed. 'I'll take Glad up a cuppa,' said Fred. 'I expect she could do with one after all she put away last night!'

When Fred entered her room, Glad was sitting on the bed, holding her head. She told him her mouth felt like the bottom of a bird cage, and that her head hurt. She had enjoyed herself but was now regretting having drunk so much.

Fred joined me at the table and before long Glad appeared. 'I don't know how you two can eat breakfast,' she grimaced. 'I couldn't face a thing. When you two leave I'm going back to bed to recover.' We cracked up. She really did look a sorry sight.

When it was time to leave, Fred paid her for our night's lodgings and we bade her farewell. As he opened the door, it was still blowing a gale and snowing, so with collars up we marched towards the lorry.

Fred filled the radiator with water, while I checked the tyres and tow-bar in case local kids had disconnected the trailer. When all was ready, we jumped up into the cab, Fred touched the starter button and the Gardner engine flew into life. We continued our journey, heading towards Blaydon, south of the River Tyne.

When we got there, two of Hardy's eight-wheeler drivers from Northfleet, Kent, were waiting to off-load. Being so far north but meeting up with drivers from home seemed strange to me. We all squeezed into the mess room to drink tea. Two of the drivers told us about some of their experiences on the road. They both had AEC lorries with single drives, and had had some hair-raising times on badly gritted roads.

Fred told them he'd been on the road so long with a

lorry and trailer that he was used to it. 'But my trailer boy here, I've scared him shitless many a time,' he teased. 'Isn't that right, Shay?'

'Yeah, you frighten the life out of me,' I admitted, laughing.

At that moment the foreman appeared at the door, and told one of Hardy's drivers to back-in under the loading bays; they were to off-load together. He told Fred he wanted his men to unload us before dinner, which was fine by us. Fred and I pulled the icy ropes off the sheets, folded them up and laid them in the passenger side of the lorry. Then Fred lined our trailer up, so as one of Hardy's drivers pulled out, Fred backed our trailer in. While they were unloading the lorry, I put the two sheets onto the trailer and tied them down so that we would be ready for the road.

In no time at all, they had us unloaded. When Fred pulled the trailer out from under the loading bay, I unhooked it, putting a large wooden scotch under the wheels to wedge it and stop it from moving. He then backed the eight-wheeler onto it. I hooked her up, connected the vacuum, then removed the scotch from the wheels.

Before leaving, we chatted for a while with the Hardy drivers. They were off to Darlington, where they would be stopping for the night. I noticed that one of the eight-wheelers had a double sticker, which meant it had two gear levers. One of them was for changing the five gears, the other was for high and low booster. It also had geared steering, which was great when you were loaded, as it enabled the driver to turn the wheel much more easily. Its disadvantage was that you had to turn it round about six times before you were on full lock. The fun of it was that it span round like a top when you let it go. You had to be careful, though. If you had your hand in the way, it could break your wrist.

When the drivers had gone, I asked Fred what AECs were like. 'They're good motors, what you call a driver's lorry – very fast. We've got an Atkinson with a Gardner engine, which is a governor's lorry. But if you fitted an AEC in an ERF or an Atkinson with double-wheel drive, the shudder from the back wheels would damage the crankshaft. That's why AECs preferably have single-drive axles.' Fred looked at me with mock impatience. 'Is there anything else you would like to know, Shay? No? Thank God for that. Let's find a telephone box and I'll ring for a return load.'

Fred got his address book out, then sorted out some change for the phone. He wasn't away long. Jumping back inside the cab, he told me that we had to load out of Sunderland – the rates were good too. The load was three large machine rollers, two for the eight-wheeler and one for the trailer.

As we drove into Sunderland I thought it looked a delightful place to live. Fred drove the lorry into a huge shed alongside the River Wear. It was quaint along the quay, with a cobbled path running alongside the tramlines. There seemed to be a lot of activity, despite the weather conditions. I noticed a fleet of Dennis lorries being loaded with one-hundredweight bags. Further up were some Leyland Cub lorries. The accent in this part of the country was very different from ours. Everyone thought I was a cockney, and I found it very difficult to understand them at times.

The quayside workers used a four-wheeled jib crane

to load the three machine rollers which were boxed in wooden crates. The rollers had just arrived from Canada, and the docks manager was most emphatic that they had to be sheeted extremely well and kept dry at all times. It was one of the easiest loads I'd ever had to rope and sheet.

As I was about to jump up inside the cab, Fred told me that I could drive the lorry to the end of the quay if I liked. I almost ran around the front of the cab to the driver's side. I was so excited. My fists were clenched tight and my inner voice was saying, 'At last.' I'd been waiting for this moment for so long. As I reached for the door handle, I could hear my heart thumping: 'At last, at last.'

As I sat in the driving seat with Fred beside me, I controlled my eagerness and listened intently to his instructions. He told me to familiarise myself with my surroundings, such as door, mirrors and the length of the vehicle. 'Remember, Shay, it's a different game altogether sitting behind the steering wheel.'

First I looked down at the pedals. The one on the left was the clutch, the middle one controlled the accelerator and the one on the right was the foot-brake. It seemed so strange to me, the throttle being near the steering box. The ratchet hand-brake was in a lovely position, lying along the floor. 'Keep your foot off the accelerator when you start the engine,' Fred reminded me.

With my right hand I reached over to the electrical fuse box, which was just below the steering column near the driver's door. There were three large switches: pilot-light, side-lights, headlights. As I pressed the pilot-light switch, it felt very tight and well-made. I wiggled the gear lever to make sure it was out of gear, then pressed the starter button, which was in the fuse box. The engine flew into life. The gearbox was back to front. Bringing the gear lever over towards me, I pushed it forward into first gear. I reached down for the ratchet hand-brake, then pulled it back and thrust it forward hard. As I released it, I kept my right foot down on the foot-brake.

There was a loud bang as Fred released the hand-brake on the near side. I checked the driving mirror to make sure all was clear behind me. As I let the clutch up, the back wheels shuddered slightly as they took the strain. I pushed the accelerator hard to the floor, my speed now being about four miles an hour. At that moment the governors of the fuel pump came in and cut the engine out so that I had no more revs. I knew then that I had to change into second gear. This was the most nerve-racking part of the exercise. I dipped the clutch and put the gear lever into neutral. I started to count from one to six, and when I reached six, I pushed the lever into second. Away I went again. Fred shouted across to me, 'Well done, Shay.' I beamed.

In no time at all I had driven to the end of the quay. I knocked her out of gear, pushing hard down on the foot-brake. The eight-wheeler's brakes made a whistling sound and the trailer let out a high-pitched, screaming noise as she came to a halt. 'You didn't do too badly for your first time, Shay,' Fred remarked as we changed positions.

I sat in the cab, thinking for a while about what I had just done. I felt so proud of myself: at the early age of eighteen, I had actually driven an eight-wheeler and trailer with the most powerful engine in Great Britain.

Before I knew what was happening, Fred had driven out of Sunderland and we were on the road again.

Fred told me we would be stopping for the night at a roadside café on the main A1, not far from Cornforth, near Spennymoor in County Durham. It was a fair old punch. The weather conditions were still appalling. At one point Fred actually managed to reach twenty miles an hour, which was quite an achievement.

It must have been about two miles after that when Fred pulled into our digs. We had stayed in better, but at least the room and sheets were spotless. There were drivers from all over the country staying there. As we sat at the table eating our evening meal, I noticed some rather rough-looking women whom people might call 'ladies of the night' sitting with drivers.

No sooner had these drivers finished their meals than they hurried upstairs with the women who were giggling and carrying on all the way to their rooms.

Fred eyed me suspiciously. 'Shay, I hope seeing and hearing all this isn't giving you any naughty ideas.'

'Give me credit,' I laughed.

'Oh, so you had thought about it then?' he replied, grinning. 'Being serious now, Shay: what usually happens is that when there're no men here who want anything to do with these women, they sit and wait for the night trunkers to change over with the day drivers, then hitch a lift to another café, where it starts all over again.

'Doesn't the café owner say anything?' I asked.

'Why should he? So long as there's no trouble and everybody's happy he'll turn a blind eye to it. After all, it's bringing money in. But if he ever gets an inkling that the wooden-tops are about, he soon ushers them all out double-quick. After all, it's his livelihood that's at stake.'

A couple of drivers soon came to join us. They told us they were from Warrington and worked for Buckley's. I mentioned I'd seen a lot of their lorries on the road, distinctive with a painting of an elephant's head on top of the cab. 'That's right, son, they're the night trunkers,' one of the men said. 'When Buckley's had that painted on, it became a well-known trade-mark.'

One of the drivers was a really big man. You could see from the size of his belly that he was an old hand in the transport game. His comrade wasn't small either; I would say he weighed around fourteen stone. They both had a terrific sense of humour, and we must have stayed there talking for at least a couple of hours. Fred reminisced with them about the old days before the war. I sat there listening intently, fascinated by their knowledge and experience.

Suddenly we heard cheers and catcalls. One of the women was doing a striptease, and this was the beginning of a riotous night that broadened my experience and left me randy as hell. I couldn't stop thinking about Rosie in Canterbury.

When we left the café the following morning it was bleak and still bitterly cold. There must have been at least twenty lorries parked up. I filled the radiator with water, then Fred started her up. None of the drivers left the park until all engines were roaring. Then it was truckers north, truckers south – as usual.

As Fred drove down the A1, the going was slow. There had been a weather announcement on the wireless, warning drivers of black ice. The roads were lethal.

Fred heeded the warning and took it really steadily. All went well until we came to a sudden halt. The traffic queue must have been about two miles long, the longest jam I had yet been in. Eventually we found out what the hold-up was. There in front was an overturned lorry. Fortunately the driver was not badly hurt, and no other parties were involved. He had slid on black ice. The eight-wheeler looked gigantic lying there. You could see all the axles, spring hangars, gearbox, prop shaft and sump. Diesel and oil were everywhere. It was chaotic.

The accident was the topic of conversation in every transport café and digs for months. Little did I realise at that time, but in years to come there would be juggernauts in pile-ups all over the country on a large scale. As we left the scene of the accident north of Darlington, I could see the stress and strain on Fred's face.

The lorry was losing traction now, and the trailer kept pushing the back end of the eight-wheeler round on the drive wheels. The roads froze over as Fred drove along. 'We'll stop at Catterick, Shay, and have a warm,' he assured me. I couldn't wait to get there. My poor old feet were numb with the cold.

After a long haul and treacherous journey, we eventually arrived at Catterick. As I jumped down from the cab, my legs ached from sitting in the same position for hours on end; the cold didn't help either. I tried to walk straight. It was a running joke that Fred called me 'Gabby Hayes' , because more often than not I wobbled after a long journey. My legs seemed to turn to jelly. But once the feeling came back I was all right. It didn't seem to affect Fred. He was like John Wayne – born in the saddle.

It was lovely sitting down relaxing at the table with a mug of tea between my hands. Fred and I shared the *Daily Mirror* as usual. The comic strip 'Jane' was popular with most drivers; she was always naked. For the first time in ages I felt warm. I didn't want to leave the café. It was a really bad year in more ways than one. A lot of good drivers were killed in the terrible weather conditions. I hoped things would improve through 1948.

In no time at all we were back on the Al, trunking south. Fred didn't like stopping for too long in case the engine froze up. If that happened, we'd be in shit street for sure.

As we neared the other side of Boroughbridge, we saw broken-down cars abandoned on the side of the road. They had almost disappeared in the snowdrifts. The only traffic on the roads now were lorries, and they were few and far between. We drivers of Atkinson lorries now lived up to our name of 'knights of the road'. Unless, of course, we saw a steam lorry driver – then the title was handed over to him.

Fred was keeping an average speed of eighteen miles per hour through towns and villages. He said, 'If all goes well, we'll stop at Newark.' As we drove through Blyth it seemed to me that the temperature had risen a couple of degrees, and I mentioned this to Fred.

'You silly bastard,' he laughed. 'What do you mean, the temperature's risen? My feet are freezing and my fingers feel as though they're going to drop off any minute. Your brain's gone.'

I retaliated by saying, 'Of all the drivers I was picked to mate with, it had to be you, Fred Ruddock – a born moaner, bloody ugly and Catholic to boot.' Fred cracked up laughing.

'I'm not amused, and if you carry on like this, you Church of England ponce, you'll be thrown out of the cab and put on a rattler home.'

Then he stopped joking, telling me he'd got a headache coming on as he put his foot down on the accelerator. As we drove into Newark, he said, 'Shay, you'll like the digs here.' I asked if there happened to be a pub up the road with a piano. 'Well, now you come to mention it . . .' he grinned.

Newark looked very clean and spacious. In the middle of the town stood a castle, which gave the place real character. I could see it was a very historic place, with lots of beautiful ancient buildings.

The guest-house had a warm, welcoming feel about it. 'I think I'm going to like it here,' I told Fred and he winked at me approvingly. As we booked in, the landlady told us it would be an extra tenpence if we both wanted a bath, which of course we did. Fred handed her a shilling, telling her to keep the change.

Once we'd bathed and shaved, we went to the dining room. The warm fire looked very welcoming. 'I don't fancy going out tonight, Fred, and leaving this lovely fire.'

'Never mind,' Shay. 'You can always stay and listen to the wireless while I go for a jug down the local.' That did it. No way was he leaving me there on my own all evening.

The landlady did us proud with a very tasty meal, and lots of it. I don't know how truck drivers would have coped without these wonderful people. They weren't like other landladies; they were a very special type of woman and they really knew how to look after hard-working transport men. If a driver stayed in a place that was excellent, word would spread, and in no time at all long-distance drivers from all over the country would have logged it into their 'night-out' diaries.

After dinner we left the cosy little haven and trudged out into the snow to the nearest pub. The problem with being a long-distance driver is that the only place you can go is a pub or cinema. You don't stay in a town long enough to do anything else. Fred's and my expenses didn't cover living it up every night, so, like most drivers, we had to fiddle to get extra money in our pockets. That was one reason why I liked working with Fred: he made it pay. Although the pub didn't have a piano, it was very pleasant sitting quietly while Fred told me more about the days he'd worked for Pickfords.

He carried on with his account of dealing with hump-backed bridges. 'First of all, we had to make sure the bridges were safe to drive over. Then we laid sleepers down on the road to try and prevent the low loader digging into the road. If you weren't careful, Shay, you could ground it quite easily and the damage would be permanent, especially when you had a load of two hundred tons or more. It's not as if you could call on a crane driver to off-load you at the drop of a hat.'

In that case, what you had to do was alternate by first jacking the front up, then the rear. If, for instance, you were pulling two hundred tons, and the lorry wasn't running on smooth ground, the solid tyres dug in and that would put stress on the engine. That was why training with Pickfords from a youngster was essential. 'If it it hadn't been for my old woman being so awkward, I would still be working there now,' Fred said.

'Surely they'd take you back.'

'Oh, they might if they needed someone. But the trouble is I've got into the habit of driving too fast. I don't fancy driving at eight miles an hour again after getting used to doing twenty.'

I'd just ordered another drink when two men walked into the pub. 'I know those two!' Fred exclaimed. 'Jock Daniels and Fred Warren – they both worked for Arnold's at Gravesend, long before the war. They could tell you a few stories about road transport, Shay.'

Jock and Fred spotted us in seconds and came over. The conversation moved to the unions. They said it was inevitable that in the near future a stronger union would be formed in the North Kent area, because most drivers had had enough. The war had brought things home to them.

The four of us had a game of darts – after a fashion, that is, because it was soon obvious that none of us was any good. The darts hit the deck and side of the board more often than not, but it was all good fun and passed a couple of hours.

As we made our way towards the lorry park the following morning, Fred told me that Arnold's had bought their first artic in 1946, a Reo Speed Wagon. The trailer had a single axle and stood high off the ground. The front of the wagon was very long; it must have been an impressive sight. Then I remembered vividly what Fred had said about their lack of power when they were heavily laden.

Jock had a six-wheel Leyland Cub, which also had a long bonnet and a low cab. They too had proved very successful in their day. Fred waved Fred and Jock on, not wanting to hold them up, as they were empty and we were fully loaded.

We punched on down the A1, driving through Blackwall tunnel, and before long we had reached Kent. As we came up onto the Rochester Way, just past Eltham, I noticed the boundary sign. The wind had blown the snow from it and I could see, above the word 'Kent', the emblem of a silver horse rearing on its hind legs – the Invicta sign. Thomas Aveling of Medway put this emblem of Kent on the front of all his steam rollers.

Near Swanscombe cutting we pulled in at the Merry Chest café which was packed as usual with friendly local drivers working for Swain, Arnold, Arthurell, Bowater, Hardy, Robarts and, the biggest of them all, Atkins of Bell Street, Medway. To Fred's offer of tea, everyone said yes in that usual good-natured way. In this good crowd of drivers, all the stress and frustration of the road disappeared and we could all relax for a while. The drivers understood each other – a special bond between them. As always, the jokes were coming thick and fast. The laughter was so loud I missed half of them.

Fred once told me that if a man had worked in a factory or office and then left, he would most likely lose touch with his workmates. But in the transport industry, it was totally different, because you would still meet your ex-colleagues in digs, cafés and wherever drivers loaded or off-loaded. This happened in all parts of the country.

Meeting in places like the Merry Chest was like being a member of a special club. For those who had retired or were unable to work any more the café brought back precious memories. Those who were unemployed could talk to other drivers who would be only too willing to put a word in for you when vacancies came up in their companies.

After an hour in the café, with tea running out of our ears, we drove on down the A2 towards Sittingbourne and Bowater's paper mill, where our next load was. It was a fairly modern industrial factory with up-to-date equipment, and we unloaded the lorry and trailer in about forty-five minutes.

All around the yard were bright red, eight-wheel AEC Mammoth Majors, all built to Bowater's specifications. Fred drove down to the far end of the factory and, noticing that nobody was about, asked me if I wanted to have a drive. He didn't have to ask me twice. I was out of that cab like a rocket and had actually opened Fred's door before he had even had a chance to move. 'Fucking hell,' he said, 'I thought your arse was on fire, you moved so quick.'

There was plenty of room for me to manoeuvre, but Fred told me to drive carefully and to do as he said. In no time at all I'd put her into second gear and moved off. Just before I put the gear lever into third, I felt the steering wheel vibrate in my hands, and was again filled with elation. I drove along in third gear for a while. Looking in the wing mirror, I thought the lorry and trailer looked longer than ever. As I approached a stack of pulp, Fred asked me to do a complete U-turn and drive back to where I'd started. I put my heart and soul into turning the lorry round. The steering was really heavy, despite the lorry being empty. Fred told me it would build my muscles up.

Turning the lorry around took up a lot of room, and I did it three times before finally getting the hang of it. I pulled over for Fred to take the wheel again. He praised me and told me I was a natural. Then it was time to head back to Banksy's.

When we arrived outside Banksy's office, he was talking to Bill Warnett. Fred went over to join them, telling me to drive the lorry over to the pump and fill her up with diesel. For a few seconds I didn't understand what he was saying. It shook me out of my shoes.

Not to be deterred, I jumped up inside the cab and sat for a while behind the steering wheel, pondering. Then, knocking the hand-brake off, I nervously moved. I could feel three sets of eyes looking at me. The girls in the office were also watching, which made the situation worse. I was pleased when the lorry didn't shudder as I drove round to the diesel pump. Luck was with me. I spotted Deafy washing a gearbox in kerosene outside the garage. He nodded his head in approval as I passed.

To fill up with diesel, I had to turn the handle backwards and forwards about 112 times just to put fifty-six gallons in the tank. My poor old arm did ache. Deafy sauntered over to me when I was checking the engine oil. 'Back the lorry and trailer into the garage, Shay, and I'll look it over while you're here,' he said.

When I shouted at the top of my voice that I wished I could, he responded with 'Eh?' He was so deaf it was exasperating at times. When he finally heard me, he said, 'At least you admit you can't back it in. Most of the drivers here can only drive forward, but they won't admit it. Just look at the state of my garage. It's got scars to prove it.' I looked and thought to myself: that's not scars, that's rust. The garage was falling to bits.

Banksy didn't say a word to me about driving the lorry, but gradually he began to ask me to shunt lorries to and fro in the yard.

The winter of 1947–8 was at an end, and I had reached the tender age of nineteen. I'd driven quite a lot by now – illegally, of course, but many drivers did in those days. Fred Ruddock had taught me well. I was a confident driver. Fred often let me drive the lorry from A to B, usually through the night, when the wooden-tops weren't around, though we still kept our eyes open, just in case.

Big companies like Sutton's were running trunk lorries; transport was really getting organised. Drivers now had change-over points. Not only were they running legally, but the lorries were in tiptop condition. They were a force to be reckoned with.

However, there were still a few of them running bent, and some companies were not so good. Their lorries had a lot of GV nines (defects) and dodgy tyres. Some drivers were so hard up they slept in their cabs.

But times gradually changed for everyone. British Road Services was becoming established, and the government was establishing pay rates, so now if a driver could not get a delivery back home, the rates still covered running home empty. So, getting a return load became an excellent bonus rather than a necessity. Most drivers didn't only do a day's driving; sometimes they had to tail-lift fifteen tons of various goods.

If you were away from home for weeks at a time, it

was embarrassing turning up at your digs looking dishevelled, dirty and a little the worse for wear. Drivers began to take more pride in themselves. Not only were the companies getting better, but so were the drivers.

Chapter 6

Big Changes

ON our way through Eltham, Fred stopped and made a telephone call to Banksy. After five minutes he climbed back into the cab blaspheming. 'That's all we need! We've got to go to Northfleet and load paper sacks – it's a right poxy job.'

We arrived on the job at twelve noon, and didn't leave until six that evening. It was bloody hard work. The paper bags were in bundles and had to be stacked on the eight-wheeler and trailer. Stacking it was a nightmare; there seemed to be no end to it. When we'd finished, it was at least ten feet high. Fred told me that we dared not go any higher because it would be dangerous.

At that moment the charge-hand appeared and shouted out sarcastically, 'Is that all you're loading?'

Fred yelled back, 'What do you mean by "all"?' He looked so angry I thought he was going to bust a gut.

'Er, nothing,' the charge-hand shouted back nervously, before walking off muttering.

'Shay! If Banksy thinks we're coming back here again he's mistaken. He can poke his eight-wheeler and trailer right up his rear, no messing.'

As always, I kept quiet when Fred was in this frame of mind. It didn't help when the sheets would not cover the load either, and we had to tuck paper inside the lorry sheets to cover the bottom of the sacks on the lorry. We then double-dollied every knot and pulled down hard onto the bags. Fred picked up the delivery notes on our way out and drove in silence back to Banksy's yard.

When we arrived Deafy took the brakes up. He worked long hours and at times I wondered if he had a home to go to. After that I checked the lorry and trailer. I could hear Fred arguing with Banksy, telling him the only work he gave us was shit. Fred was in a foul mood. I'd never seen him like this before.

Banksy retaliated: 'Most of the work you get is away from home, and you always make sure that it's a crane load, on and off. So a dodgy load won't hurt you now and again. What do you think you are anyway, a bloody chauffeur?'

Before I knew what was happening, Fred had grabbed Banksy by his lapels and was about to punch his lights out. I ran over and caught hold of Fred's arm but he shrugged me off.

'I was only joking, old boy,' Banksy choked. 'You're tired. Go home and get some rest.'

Fred stepped back, turned and walked away, muttering, 'Toss-pot!'

That night I went back with Fred to his digs in Brent Road. His landlady was all right about it although, as I told Fred, she looked a miserable old cow.

'I've known better, Shay. But the rooms are okay and the price isn't too bad. When I'm ready I'd like to buy my own house.'

'Sounds good to me, Fred. You'll be able to put me up for the night when I'm on the road,' I replied cheekily.

'I'm going to start the way I mean to carry on, Shay. That means no riffraff. Sorry.'

'Oh, and I thought you'd cheered up!'

'I have, don't you take any notice of me.'

Fred and I were on the road by five the following morning. The weather was a bit brighter, and it wasn't too bad driving through London alongside the embankment.

All of a sudden, a night trunker coming in the opposite direction gave Fred the headlight sign. Fred pulled over and stopped. When we jumped down from the cab, we saw all the ropes dangling, and the lorry sheets flapping about. 'Bloody hell,' I said. 'What's happened here?'

'All the air's come out of the bags, which makes the load shrink down with the weight,' Fred answered. Our load was now eight feet high instead of ten.

As Fred and I were roping, a couple of drivers from Yiddle Davis in London pulled up behind us. They could see what had happened, and gave us a hand to re-rope the lorry and trailer, which made them late because the day drivers were waiting for them to change over. We thanked them for their trouble and were on our way once more.

In a short while Fred was driving past Earls Court and out of London, heading towards the A40, going west. We stopped for breakfast on the way at the Towers café and were able to spend an hour in there. Despite the hold-up, Fred and I had done well that morning.

Fred told me as it would be a fair old punch we would not be stopping any more until we reached Bishop's Cleeve. We seemed to be climbing up more hills than driving down them. The A40 was very narrow and as we drove through some of the villages, I noticed that many of the bedrooms of the houses jutted out into the roadway. A thought suddenly struck me: if Fred drove a bit slower, I could almost stretch out and tap on one of the windows for a giggle.

As Fred drove into Cheltenham, I spotted a small café on the right-hand side. Fred caught me looking at it and said it was a good place to eat. The owner also owned digs around the corner from the café, which was convenient for drivers, and both the café and digs were excellent value for money.

I thought Cheltenham was a splendid town. The architecture was unequalled by any I'd ever seen, and with its matching buildings surrounded by iron railings, the place was very conservative looking.

Fred drove around the one-way system and headed towards the A435, eventually arriving at the factory, where we reported to the Goods In department. The man there asked us to take the ropes and sheets off our lorry. He told us that they didn't often get lorries with trailers. Fred backed the lorry in. He certainly had the knack of backing trailers, and some of the workers watched in admiration. They themselves were good workers and soon had us unloaded. Fred and I stayed and had dinner with them in their canteen, which wasn't too bad – it saved us a few pennies too.

After many phone calls Fred managed to get a return load of fire pumps because, as luck would have it, we had a trailer. The company wasn't interested in the weight we could carry but the amount of space we had. Tewkesbury Transport gave us the load and they assured Fred that their rates of pay were good and conformed to the Ministry of Defence pay scale.

We had to report to the Royal Engineers' sub-depot, Honeybourne, just outside Evesham. We were loaded by four drivers who drove four massive six-wheel Albions with large cranes on them. I could not help laughing; the Dennis fire pumps only weighed about fifteen hundredweight each, and I thought the army was going over the top a bit. They loaded five sideways onto the eight-wheeler and three sideways onto the trailer. Fred and I didn't sheet them, we just double-dollied them down with ropes.

An officer in charge strolled over to us and spoke in a highfalutin' voice. 'By Jove, you've got a lot of pumps on board there, old chap.'

We had to laugh. If they had put a one hundredweight bag on a three-ton Bedford, to them it would be overloaded. Much to my embarrassment, Fred asked the officer if we could stay there for the night. 'No trouble at all, old boy. No trouble at all.' Fred offered to pay but he wouldn't hear of such a thing.

The officer showed Fred where he could park. I noticed quite a few army lorries there. Some of the sappers – roughly my age and a bit older – gathered round to look at our vehicle more closely. They took us over to their canteen – it was the first time I had ever eaten two dinners in one day! One of them asked me why I was not in the army, and I told them it was because I'd got a crooked spine and the army wouldn't have me, much to my relief.

After we'd finished our dinner, the sapper they called Jimbo showed us to a Nissen hut where we were to sleep that night. It was very comfortable.

We woke the following morning to the sound of 'Quick march' and 'Left, right, left, right.'

'Bloody hell, Fred,' I laughed. 'I'm glad they didn't take me into the army. I don't think I could have handled all that!'

Fred laughed his head off, telling me it would have 'done you good, Old Chap'.

Fred started to make out the log sheets for that day. 'I'll book us on at four in the morning and last night I booked us off at Sedgeberrow. By doing that, it looks as though we've done five hours work already, especially booking loading today. On paper we will leave Honeybourne at nine o'clock.'

I chuckled to myself as Fred drove out of the main gate. 'He's a lad,' I thought. 'A right old crook.'

We were now headed for Burscough in Lancashire, on the other side of Liverpool. Apparently the pumps we had on board had to be repaired, so it looked as though we'd be reporting to another army base. 'That means, unfortunately,' said Fred, looking very pleased with himself, 'that our eleven hours will be up by the time we reach Warrington.'

'Warrington!' I shouted happily. 'Brilliant! That means we'll be seeing Lancs.'

Fred rested his hand on his chin and said, 'I didn't think of that, Shay.'

'You crafty old bastard!' I replied. 'That's why you've been fiddling the log sheets.'

Fred headed straight towards Evesham, where we picked up the main road for Birmingham. Fred drove well over the speed limit, the engine roaring and the poor old lorry unable to go any faster. I kept a keen eye out for the wooden-tops. Fred knew Birmingham well, and before long we were travelling along the A34, heading towards Cannock. He asked me if I wanted to stop, but I told him to keep on tramping, as I just wanted to get there.

As soon as we had passed through Newcastle-under-Lyme, Fred yelled across to me, 'Engine room, Shay!' I immediately lifted the side panel off the engine bonnet. As I looked down at the front axle, the road was flying past underneath us. The engine looked larger than normal and the noise was deafening. I bent down and pulled the rack back with my right hand, and as I did she chucked out black smoke, though only for a short while before settling down.

Fred was really motoring at 42 mph. He kept the same speed, whether driving up hill or down dale. He even drove through a couple of villages, still holding her at the same speed. I was getting a little concerned in case the law pulled us up, as speed traps had now been set up and the police were keen to try out their new toys. Also, they would more often than not pick on lorry drivers as they were an easy target. I suppose it made their little notepads look good.

Our plan was to go along the A50 and make our way north towards Knutsford. Having only two lanes made it difficult to overtake other vehicles. Now and again there was an overtaking lane; Fred would take advantage of that and overtake anyone who got in his way. He was in one of those moods again. 'I expect I'll be just the same when I eventually drive,' I thought, but he still made me nervous at times.

The A50 was a strange road, as one minute there was no traffic at all, and the next you were sat in a queue. But the condition of the road was a lot better than the dart board on the A6. Fred shouted across to me: 'Let the bar go now, Shay', which I did, at the same time putting the engine cover back on and replacing the blankets, making it a lot quieter inside the cab.

Fred stopped when we reached Warrington. While he was fiddling the log sheets, I lifted the side of the engine cover up again and checked the oil. Finding it was low after having had the bar back, I jumped down from the cab, took the key, opened the toolbox that was on the side of the eight-wheeler, retrieved the can and put four pints in. After doing so I put the can back and locked up.

'I've booked eleven hours today, Shay, although we've only worked six. We gave a bloody good day's work yesterday. Now we'll go and see the old landlady and book our digs.'

As we were early and not far from the Red Brick pub, I offered to treat Fred to a pint before they closed for dinner. 'I never say no to a pint, Shay.' On reaching the pub, I walked up to the bar and ordered us two pints. I saw by the look on Madge's face that she was pleased to see us. Madge told us that they were retiring and moving back to Leyland within the next couple of days or so. I asked if Joyce was going with them, and her reply was a definite yes, as she'd got a job as a clerical assistant there.

'Joyce will be so thrilled to see you both again,' said Madge. 'I'll give her a shout. We've been wondering what had happened to you both.' She thought Fred might have retired from road transport, and I told her we had been very busy.

When Joyce emerged from the back, her face lit up when she saw Fred and me standing there. I thought she looked as beautiful as ever. Fred thought so too, I could see that. He didn't take his eyes off her once. In a soft voice she said, 'Hello' and asked us how we were.

Fred told her it was nice speaking to her again, and she replied, 'Likewise.'

'I've heard there's a park here,' I remarked.

'Yes, there is, she said, and it's lovely.

'Fred was wondering if you would show him round the town and then perhaps go for a walk in the park.'

Fred glared at me. 'I can speak for myself, Shay.'

'Then why don't you?' I replied, winking at Madge.

'Some fresh air would do her good,' Joyce's mum replied with an amused smile. Finally Fred asked Joyce if three-thirty would be all right, and Joyce said that would be fine by her.

As we left the pub, I mouthed to Fred, 'We've done it!'

'Thanks for that, Shay. You're a good lad.'

'I just felt sorry for her. I didn't think you were ever going to ask her out, you miserable old sod.'

'Hey, not so much of the old, Shay!'

'At least we agree on something, though,' I smirked.

'What's that then?'

'You're a misery.'

When we arrived at the digs, the landlady looked pleased to see us again. As Fred was sprucing himself up, he asked me for some ideas as to where he could take Joyce after the park. 'You can always hop on a bus into Manchester and have a meal there. I'm sure you'll think of something to do. Oh, and by the way, don't forget to tell the landlady you won't be in for dinner.'

When Fred asked me what I planned to do, I replied, 'Don't worry about me. You go and enjoy yourself with Lancs. I'll find my own amusement in the Red Brick or go to the Dollies.'

As Fred walked to the door, he turned and said, 'Do I look all right Shay?' He looked as bloody handsome as ever, and I told him so.

After about three hours I became bored, and so was thankful to sit down to my evening meal. Afterwards I read the local paper, then finally decided to give the local cinema a try. The landlady told me a Western was showing called 'Duel in the Sun', starring Jennifer Jones and Gregory Peck.

When I came out of the cinema it was still a pleasant evening, so I strolled up to the Red Brick for a pint. Inside the pub there were a good few people sitting round talking. As I ordered a pint, the landlord said, 'Madge and I don't know where we are, what with all the packing and trying to serve at the same time.'

'I'm at a loose end, Pete. Can I help in any way?'

'I'd be very grateful, if you wouldn't mind.' So I started to pack their crockery and glassware in a tea chest, being careful not to break anything. While I was packing, Pete remarked that I looked like a fish out of water without Fred. I had to chuckle to myself, as I was doing fine while both he and Madge were flapping around like blue-arsed flies. They really did make a smashing couple.

A few more customers came in later that evening. Pete served them and left Madge and I to it. Madge kept plying me with beer. Every time I emptied my glass, she'd fill it up again and, being a man of Kent, where the hops were grown, I never refused.

When the packing was done, I asked what else I could do. 'Can you clear the garden shed for us?' When I opened the door of the shed I saw that it was crammed full. Brass candlesticks stood on the shelves. There were old pictures, even photographs in frames all piled up – everything bar the kitchen sink.

I asked Madge if she wanted to keep anything. She handed me a bag: 'Clear the lot out, Shay. If anything's any use you can take it. Most of it has been left there by the previous occupants. Pete and I never seem to have the time to clear it out.'

I didn't waste much time and within an hour I'd put it all into the bags ready for the dustman to take. In the end they'd given me so much to do that I was beginning to wish I hadn't volunteered, but at least it kept my mind occupied, and they really appreciated the help. Madge invited me to stay for supper after they closed, which I did and thoroughly enjoyed it.

Madge reminded me of my old mum, trying to glean as much information about Fred as possible, but doing it in a very subtle way. I told her I thought the world of him and really enjoyed working with him. Madge told me she shouldn't mention this, and didn't like speaking ill of the dead, (which amused me, because she intended to carry on anyway), but Joyce's marriage hadn't been a good one. Apparently he'd been a womaniser, though what she was saying was between her and me and she didn't want the conversation to go any further. I told her Fred was the opposite of that. 'I drew that conclusion from the start,' she replied. 'Pete and I like him very much.'

Just as we had finished talking, we heard the sound of a key being turned in the back door. 'That'll be our Joyce, probably with Fred,' Madge remarked. She asked them if they'd had a nice time, then went into the kitchen and put on the kettle. I could see by their faces that they'd enjoyed themselves. As Madge came back with the tea, Fred remarked, 'I hope Shay's been behaving himself.'

'He's been a tremendous help to us, Fred. A real brick.'

Pete told us that there was still a lot of work to be done, and that it was hard to pack and serve customers at the same time. He said that women had the knack of doing two things at once, but he couldn't and got himself into a right old state.

'Would you like Shay and I to give you a hand all day tomorrow?' Fred offered.

'That would be kind of you, but what about your own work?'

'Oh, it won't take us long to get to Burscough and unload. If we start early, we can be back by ten. We've been working our socks off without a break for weeks. We both need a rest.'

As Lancs walked us to the door she thanked Fred for volunteering to help her dad, saying he was coming up for his sixty-fifth birthday in a few months' time, and she was getting concerned about his health. Fred told her not to worry about a thing. She gave him a quick peck on the cheek and disappeared inside the pub.

'I hope the landlady doesn't hear us, Fred. We don't want to disturb her. It's very late.'

But then as he turned the key in the lock, she shouted out to us, 'Do you two lads want a cuppa?'

'Don't mind if we do,' we replied together, sounding very harmonious. 'You haven't been waiting up for us, have you?'

'No, not at all,' said our kindly landlady. 'I was listening to the wireless and doing some knitting. There are some good plays on late at night. I like the murder mysteries best.' We drank our tea, then retired to bed, telling her we had to leave bright and early in the morning, and would be as quiet as dormice.

We woke early the following day, and crept out quietly. As we walked to the lorry, my tummy started to rumble. It was the first time I'd started work without having breakfast first. Fred hadn't even mentioned food or drink, and I thought to myself: 'It must be love.'

Fred drove along the A49, heading towards Standish. When I asked Fred the time, he said it was half-past-five. 'What!' I exclaimed.

'You heard! I can see I'll have to nickname you Deafy number two.' Before I knew what was happening, my head was resting on the bonnet and I'd fallen fast asleep.

As Fred drove along, I woke up and noticed a signpost which read A5209. 'Oh, Sleeping Beauty has woken up, has she?' I apologised for being such a wimp, but Fred said not to worry about it. We would soon be in Burscough.

After twenty minutes Fred pulled into the main gate which was manned by two civilians. There were also army personnel in the yard. One of them told us that no-one was around yet and there wouldn't be any movement until about seven-thirty, but the army canteen would be open. He said if we made eyes at the canteen staff they would give us breakfast.

We thanked the civvies, walked off towards the canteen and joined the queue along with the army lads. The breakfast looked appetising, but then anything would have; I was bloody ravenous. As Fred and I were just about to sit down, I exclaimed, 'How are we supposed to eat this? We haven't got any eating irons.'

'Oh, I'll go and ask that sergeant over there where I can get some. He looks full of piss and importance,' said Fred, none too quietly.

I cringed with embarrassment as I'm sure some of the men had heard. Fred just didn't care what people thought of him. The duty sergeant borrowed some cutlery from the cooks. He told us that they 'had backs to them' by which he meant they were just on loan to us and had to be returned.

It was a fair old breakfast. Then, much to my astonishment Fred announced that he was going back for seconds. I bowed my head, thinking: 'He's not with me.' The duty sergeant didn't say a word. Just as we had finished eating, one of the security men walked over to us and said we could unload the pumps ourselves, otherwise we might be there for at least another hour.

In no time at all Fred had backed the lorry onto the loading bay, unhooked the trailer, and backed the eight-wheeler alongside it. We jumped up onto the trailer and while Fred guided the pumps, pulling at the same time, I pushed from the back. Within half an hour we had off-loaded them. Fred asked me to back the eight-wheeler onto the trailer, while he connected it up.

Without saying a word, he then jumped up inside the cab on the passenger side and I immediately jumped up behind the steering wheel. Fred instructed me to drive to the main gate, which I did without making any balls-ups. I felt very pleased with myself. The 'utility copper' signed our delivery notes at the gate on our way out.

'Do you feel sure you're ready, Shay, to drive an eight-wheeler and trailer on the public highway?' I didn't really know, but Fred put my mind at ease by saying he thought I was. I eased the lorry into gear and moved off slowly. On reaching the end of the road, I turned the steering wheel round hard, stopping at the main road. When the road was clear I looked through the nearside mirror and pulled away. Seeing that I was at a right angle, I straightened her up. Moving up into fifth gear, I held her back to twenty miles an hour. I was in a complete world of my own, concentrating hard on my driving.

Coming in the opposite direction was a four-wheel ERF with one of Smillie's drivers from Glasgow at the wheel. The badge on the front had been taken off the radiator grill and replaced by a large 'V.' He acknowledged us as he drove past.

I didn't realise it at the time, but I was exaggerating my head movements by checking both mirrors every few moments. I suppose you could call it being over-cautious! Fred piped up, 'Shay, you look just like a spectator watching a tennis match.'

On the outskirts of Parbold, I cautiously slowed up and changed down through the gears. As I turned into a right-hand bend, the trailer clipped the kerb, which unnerved me a little and I started to flap. Fred put me at my ease again. He was a very patient person when he wanted to be. As I drove into Standish, Fred asked me to pull into a lay-by when it was safe. After a few yards down the road I did so, pulling the ratchet hand-brake up. 'How did I do, Fred?'

'Not bad at all. Well done.'

'It worried me a bit when I hit the kerb.'

'That's the least of your worries, Shay. Everyone does that from time to time.'

As soon as Fred jumped up behind the wheel, the lorry seemed to come alive. It responded to him in every way, as though it was a living thing instead of a man-made machine. This is a feeling that I cannot explain. Only other drivers would understand.

Soon we were travelling along the A49 towards Warrington. Fred parked the lorry not far from the Red Brick where Pete and Madge were pleased to see us. Madge told us all the packing was done, and that Pete had got all concerned over nothing. She apologised to us profusely for causing us any inconvenience.

Fred used her telephone to ring Buckley's of Warrington who told him they wanted a load trans-shipped from one of their lorries. Madge didn't feel so bad when Fred told her we wouldn't be losing a day's work after all.

It didn't take long to reach Buckley's yard, where Fred reported to the transport office. They asked him to pull into the middle of the yard. Within no time at all an eight-wheel Scammell had pulled alongside our vehicle, and a four wheel ERF had backed onto our trailer. We had plenty of help trans-shipping from one vehicle to another; drivers just seemed to appear from nowhere. One of them Fred knew from years ago.

We had plenty of help with the roping and sheeting too – they were good lads at Buckley's, very co-operative. Fred asked one of them if it was a dodgy drop, and why their own lads were not delivering the loads.

'No, it's nothing like that at all. It's because we have so much work on. We've never been so busy, and these two lorries we're off-loading will soon be loaded with something else.'

The transport manager gave us an eighteen-ton cap payload, although we only had fourteen tons in weight on. The bags were large but light. The manager told us we could deliver any time the next day, which suited us.

It was one-thirty in the afternoon. It had been raining quite hard earlier, but now the sun had just started to break through, which made a pleasant change from all that snow we had endured. Fred pulled onto waste ground, about two hundred yards past the Red Brick.

'We'll stay here for tonight, Shay. I'll book us off at Newcastle-under-Lyme to give us a twelve-and-a-half hour spread. Tomorrow we'll have to make up six hours work. But in the transport game, you get your hours in first!'

Since it was early afternoon the landlady was surprised to see us when we came to book in for the night. Fred asked her if it was all right to wash and change and forgo the evening meal, adding that she wouldn't lose by it, as we'd still pay the going rate. 'Oh, I don't know, one minute it's no breakfast, now it's no evening meal! What next? I ask myself,' she quipped.

'That's nothing. Wait till you hear what time we're starting in the morning!' Fred grinned.

'Surprise me,' she answered.

'We'll be leaving at six.'

'Oh, that's no problem, I'll make sure you have a hearty breakfast before you leave.'

'You're an angel, one of the best,' Fred told her.

As we came downstairs, we shouted cheerio to her. She came out of her room and handed Fred a key, telling us with a smile to enjoy ourselves and not do anything she wouldn't.

No sooner had we stepped outside the front door

than she piped up: 'Give my love to Joyce,' which stopped us in our tracks. I returned her smile and told her we would. As we walked through the front gate, Fred commented that it was typical of a small town that everyone knew everyone's business, but no harm was meant by it.

In the Red Brick, Fred and I spotted a new face pulling pints behind the bar. We guessed it was the new landlord, and Pete was showing him the ropes. As we moseyed towards the bar, Pete automatically poured out two pints of best for us, at the same time calling to Joyce. When she saw us her face lit up with a radiant smile. She said hello in that husky voice of hers, which made me melt, never mind Fred.

'Dad, would you mind if I went out tonight?' she said to her father.

Winking, he replied, 'No, you go and enjoy yourself, love. I've got plenty of help behind the bar.'

So she linked arms with Fred and me, saying, 'Come on, you two, let's go.' I protested, insisting the two of them went out on their own, without me tagging along playing gooseberry.

'Don't be daft!' Fred and Joyce replied in unison. They wouldn't hear of such a thing, and anyway, it was our last night in Warrington. I was coming and that was final.

The three of us caught the bus into Manchester, and booked a cabaret show that starred George Formby. It turned out to be A1. We all really enjoyed ourselves and it made a change from sitting in a pub, especially for Joyce. Afterwards we went to a restaurant, which was excellent.

Joyce told us she had a bungalow near Wade Hall,

Leyland, and asked Fred and me if we would like to visit her some time in the future. We told her we'd love to. Then, still chatting, the three of us strolled to the bus shelter. The pubs and cinemas had turned out, so when the bus pulled up it was pretty crowded. We finally got out in Warrington. I told Fred I'd see him later back at the digs, kissed Joyce on the cheek, turned and disappeared into the darkness.

Fred woke me the following morning, saying, 'Shay, we'll stop for breakfast on the road. It will make up for the time we've lost.' As I looked out of the window the clouds were scudding across the sky. Dawn was just beginning to break. The Gardner seemed to love this time of the morning when the air was cool and fresh, and she barked beautifully. Fred made her tramp on up the road as we made our way towards the A50.

We were headed for Corby on the other side of Market Harborough. From Warrington to Corby was about 120 miles, a good six and a half hours' driving if we were lucky. We planned to carry on down the A5 until we reached Tubbies around mid-morning, where we'd stop for breakfast.

Although the A5 was extremely busy we made good progress until we reached the other side of Atherstone where Fred completely lost all acceleration. He pumped the throttle, but to no avail, and pulled her over to the side of the road when it was safe.

Fred shouted over to me to lift up my side of the bonnet. When he pushed down on the accelerator pedal again, everything appeared to me to be normal, but when Fred lifted up his side, he saw immediately what the problem was. The ball joint had gone on the throttle linkage.

'I'll soon have it fixed, Shay. I'll tie the bastard down until we get home if I have to.' He asked me to look round for a piece of wire and along the road I spotted a wire fence. No sooner had I started to walk over to it than a six-wheel ERF pulled in. The driver was from Alison's of Dundee.

'What's wrong, laddie?' he asked.

'We need some wire. Can you help us out?'

'No problem, laddie,' he answered, handing me a brand new nine-inch reel of electrical wire.

As I was about to walk away he shouted, 'Hey, you'll want these, won't you?' He was holding out a pair of pliers.

I thanked him and ran to give them to Fred, who was amazed. He hadn't seen a thing because his head had been stuck under the bonnet.

As he tied the throttle down, another lorry pulled in front. This time it was a Mytton Mills eight-wheeler and trailer. The driver asked Fred if he was okay, and he said we were fine now, but thanked him for stopping anyway.

The Alison's driver let us keep the wire because he'd just made a delivery to the cable works and cadged some off one of the lads. It was a relief to be mobile again.

Eventually Fred pulled in at Tubbies, where we jenked up. The bloke on the forecourt put thirty-five gallons in. Fred asked him to book forty, which he did. 'That'll pay for my tea and breakfast,' Fred said. 'Don't know what you're going to do, though, Shay.'

The fellow on the pump heard and said, 'You can book a new tyre if you want!'

Fred told me to park up while he got the teas in. The

only trouble was, no sooner had I let the hand-brake off than I was winding it on again. Looking over to the passenger side, I realised the trailer brake was still on. 'Oh shit,' I muttered, hoping Fred hadn't heard anything. It was a good job I hadn't wound it on too hard.

When I got to the café, Fred asked casually, 'Everything go all right, Shay? No problems?'

'Er, no,' I stammered.

'Lying little git. You left the hand-brake on. I heard it.' He didn't miss a trick.

One of Yiddle Davis's drivers standing behind us said, 'I don't know, these trailer boys are all the same, bloody hopeless. Governors employ anybody these days. They just can't get the staff.' I felt such an idiot until I realised he was only joking.

The young lad who was serving behind the counter looked a right cocky bastard, standing there chewing gum. I took an instant dislike to him. He wrote our order down on the back of what looked like a raffle ticket: not very professional. At that moment the telephone rang. When he returned he apologised, saying that a driver wanted to book a bed for the night. 'Right, that will be two shillings and ninepence please,' he told Fred.

Fred's face went from white to crimson. 'What are you talking about? I gave you a pound before the fucking phone rang. You gave me the numbers for the breakfast, but no change.'

The Yiddle Davis driver nodded, saying, 'That's right, sonny, I saw him give you a pound.' The young man look puzzled but handed the change to Fred.

When the Yiddle Davis driver joined us at our table,

I saw Fred slide a dollar across to him and, without a word, the driver slipped it into his pocket. 'It's great eating free grub and getting paid for it, Shay,' Fred winked.

'You really had me fooled back there, Fred,' I said, a bit shocked. 'I thought you were being serious from the look on your face. Nobody would have thought you were lying through your teeth.'

'Have you just come off the banana boat or something?' commented the Yiddle Davis driver.

Fred looked at me laughing. 'You've still got a long way to go, Shay, but you're learning fast.'

As we sat talking, the driver told us about some of the pranks he and some of the other drivers got up to. Some of them really astonished me. It seemed that if you worked for Yiddle Davis there was never a dull moment, and Yiddle was very fair to his drivers. He never questioned what they did and whatever extras they booked, they always got in their wages.

After a chatty breakfast, we were soon heading towards Lutterworth, where we turned off and followed the signs to Market Harborough. 'Can you stop a minute, Fred? I'm busting for a leak.'

There was a portakabin in a lay-by, so we stopped there. As I walked in, two men were already standing at the urinal. As I stood having a pee, I couldn't help noticing that one was touching the other.

Fred told me not to be too shocked. 'It's always gone on right through the ages,' he said, 'but since the war it's become more blatant.' Soon the signs came up for Corby. Fred stopped and asked a man the way to the Bitumastic plant, which was just outside the town.

The Bitumastic plant yard looked a right old mess. In

fact it was a bloody disgrace. On the ground lay old pipes and lagging, and steam was bellowing from the pipes. Everywhere I looked there were cans full to the brim with bitumen. Fred went to make enquiries. The bloke who appeared had at least a month's growth of beard; it was all shaggy and bedraggled. He needed a good shower. He told us to unload where we were, which didn't surprise us in the least, and, to top everything, we had to unload it ourselves. Fred handed him our ticket, he signed it and ambled back to his office – if you could call it that. It was more like Aladdin's cave.

Fred and I were a real team. It took us about forty minutes in all to off-load the lot – not bad going. We were glad the job was over quickly, because the place really was a filthy hole.

Fred made a quick phone call to Kempstone in Bedford, who told him to get there as soon as possible because they wanted a full load of bricks delivered to Erith. Fred put his foot down and we seemed to fly along the A6.

Soon we were in the brickfields being loaded. The men there really handled those bricks professionally and within an hour and a half we had loaded the lorry and trailer. My hands had soon become red and sore, and by the end they were almost raw. Fred's hands had toughened over the years and mine did eventually.

We only sheeted the back end of the eight-wheeler and trailer, but roped the lorry sheet on, because that's all you had to do with bricks.

Fred drove off and turned right onto the A5120. It was a long uphill drag to Ampthill. Fred went through the gears very quickly until he reached crawler gear. I shouted across to him, 'No more left, Fred.'

'You're wrong there, Shay.'

'Why's that then?' I asked.

'We've still got reverse.'

Coming in the opposite direction there were convoys of London Brick lorries, all eight-wheelers, Leylands and AECs. They all had the same hod-carrier badge on the front of their cabs. They waved to us as they passed. Fred was now following the signs to Redbourn, passing through Markyate on the way.

The café in Redbourn made a pleasant change from Markyate. It looked a respectable place, full of transport men who were also staying for the night. As we sat drinking a very welcome mug of tea, Fred told me that as we were ahead of schedule, he would spread our time over twelve and a half hours, which meant we could have a lie-in the following morning. Then he put his hand in his pocket and handed me seven shillings and sixpence. 'That's for diesel,' he said, 'and breakfast at Tubbies, of course. Fifty-fifty, okay?' I'd never had it so good.

A couple of Anglo Scott's drivers walked over to us. One had a broad Scottish accent and his mate was a cockney. They were a great pair, born comedians, though I found the Scottish driver difficult to understand half the time. They told us that drivers who were lucky enough to be taken on by Anglo Scott had to sit a driving test first. Many drivers envied them because it was an excellent company to work for, with top-notch pay.

In some companies, drivers who wanted to run legal were known as troublemakers or just bloody lazy, so the company would get rid of them one way or another. It was extra hard for the ones who did run legal, such as Sutton's, Pollock's, Arnold's and Swain's, to name but a few. Because drivers had to buy good rolling stock, overalls, etc., the rates had to be kept high.

Every so often the Ministry of Transport did silent checks on drivers to try and catch up with those who were working bent. Drivers who had been away for two nights or more would come home on a dodgy night out on their last delivery. A few transport managers turned a blind eye to this, especially if the driver was a good 'tramper'.

The day-and-night companies would try to undercut the big ones, although it was virtually impossible to do this. Smaller firms had lower standards, making the drivers' work much harder. Some would have to tie up their exhausts with wire, and drive along with holes in the floorboards. You could actually fall through them and break a leg, especially when loading or off-loading at the back of the vehicle. This happened many times in all parts of the country.

Some drivers were so badly paid that they spent many a night sleeping in their cabs with no bedding. They'd just curl up on the seat and rest their head on the bonnet. It became known as 'the cab hotel'. These Johnny-come-latelys hadn't a clue about road transport. There were also individuals who started up in business for themselves but had little or no regard for rules and regulations. It was beyond me how they managed to get away with it for so long. But when the law finally caught up with them they became bankrupt overnight.

Drivers were getting return loads, but their lorries did not carry an 'A' licence, which allowed a lorry to be driven anywhere, and return with a load. There were some lorries which only had a 'B' licence being driven all over England and Scotland whereas officially such vehicles could only work within a twenty-five mile radius of the driver's depot of registration. It was certainly hard for the big companies that ran legal.

The 'C' licence allowed a lorry to be driven anywhere in the country but it could only deliver goods that were contracted out. It was also compulsory to run the lorry home empty, so companies that had 'C' licences tried their utmost to get an 'A' Licence for their lorries.

One day Fred suggested we take a walk, as he wanted to have a chat. That sounds ominous, I thought. It was unusual for Fred to want to walk. He hardly ever did, though it was good exercise for the likes of us who sat on our rears all day.

'What did you want to talk to me about, Fred?' I asked.

His voice wavered a bit as he told me he was going to take a fortnight's holiday and would tell Banksy when we arrived back at the depot. He didn't give a tinker's cuss about Banksy not having anyone to take his place – that was Banksy's problem.

Fred said that he hadn't taken a break for two and a half years, and it was about time he did. 'Now I come to think of it, Shay, you haven't had a holiday either since you've been working with me. It must be at least two years.'

'That's true, Fred. But the money's come in handy.'

'To be honest with you,' Fred confided, 'I'm completely pissed off now with work, work, work. The money hasn't been too bad, fair enough, but it's not everything. And I've nowhere permanent to live. It would be nice to settle down.'

We both fell silent for a while until I broke the silence by asking with a smile: 'Your place of abode wouldn't be Leyland, with Lancs in mind, would it?'

He looked me straight in the eye. 'I think you've got it about right there – assuming she wants the same, of course. Joyce and I have talked about it. We know we've only known each other a short while but we're not getting any younger, so I think we know our own minds. It's funny, but we hit it off right from the start – mutual attraction, I suppose. Keep what I'm saying under your hat, Shay. I don't want all and sundry knowing at the moment.'

'Mum's the word,' I promised.

'If the future holds good for both of us, she'll sell her bungalow and I have the equivalent in savings, so we could buy a bigger house somewhere. Maybe Leyland, who knows? But after all that, I've got to be sure and so has Joyce. Once bitten twice shy, as they say. Eh, Shay?'

I knew I was being selfish but I kept thinking: what will happen to me? Banksy had only got one trailer and they would never get anyone else willing to work away as much as Fred did. And if I worked locally on a wagon and trailer, I'd just be a lackey, humping bags all day. But Fred reassured me. 'I'll stay on at Bansky's for at least another year or so, Shay. Don't worry your head about anything. I'll help you get your licence.'

I was so relieved. 'Is that a promise, Fred?'

'I always keep my word. You should know that by now. But I will say this,' he added, grinning, 'you'll be away from home a great deal.'

'Lots of nights out in Lancashire, I suppose,' I laughed.

We must have walked at least three miles, so we turned and made our way back, as by now darkness had fallen. Although I was pleased for Fred, I also felt sad and apprehensive, knowing I was going to lose a way of life I had come to love. But c'est la vie.

Back at the digs, Fred sat up in bed rubbing his hands together with glee. 'Shay, we've got another twelve and a half hours in today!'

'I thought we were only allowed to do two spread-overs a week, and the rest elevens?'

'I'm booking twelve and a half on our time sheets, but our log sheets say eleven hours. Got it?' He was a crafty bastard, was Fred.

The following day we had a good lie-in and ate a hearty breakfast before setting off again. Off the cuff Fred booked us on at 4 am but we actually left the lorry park at eight. The weather was bright but breezy that morning, and the clouds looked like tufts of cotton wool in the sky. As we made our way towards the A5, I questioned Fred about booking on at four, saying we'd be at the customer's too early, and wouldn't Banksy query it?

'Do I have to keep on telling you, Shay. I'm the captain of this ship. I want to be finished at four o'clock this afternoon with eleven hours' pay. But if I listened to you we'd still be working at eight o'clock tonight for our eleven hours. It will only take us about three-quarters of an hour to drive back to the depot. If we're early we can always busy ourselves in the yard.'

'Point taken,' I conceded.

The bricks we carried were low down, making the

load look very square. The vision was good too, compared with some loads. For a minute I sat there, watching the trailer rock slightly to and fro. Then I decided to polish the cab. The aroma of lavender blended in with the varnished timbers.

In no time at all we were travelling through the London Colony which was north of London half way between St. Albans and South Mimms. I had got to know the area well by now and I knew shortly we would be driving through South Mimms, then onto Barnet, Archway hill, Hackney, the Blackwall tunnel, eventually ending up on the A2. We passed the famous Dover Patrol pub near Eltham. Soon I spotted the Invicta sign of Kent, and knew it wouldn't be too long before I reached home.

On the approach to Dartford, Fred turned off left, heading towards Erith. We had soon off-loaded our bricks, and although it was hard work we had finished in an hour and a half. When we passed through Belvedere it was about nine in the morning.

'As we're a bit early, Shay,' said Fred, 'we'll make our way back to the A2 and stop at the Ark café in Bexley. I haven't been there for a long while.' We had a bit of a squeeze getting into the lorry park which was very small compared to others. There were a few Bowater lorries parked there.

In the café I looked around for my old man, but one of Bowaters drivers told me he had a delivery at the Daily Mirror and would be in later on. The Bowater drivers appeared to be more serious than other drivers I had met. It wasn't until Swain's and Arnold's drivers began to pull in that the jokes started to come thick and fast. The old girl behind the counter pretended to cover her ears, but even she had to laugh, taking it all in good fun when they teased her.

We were in stitches as we made our way back to Banksy's yard.

Fred pulled round onto the diesel pump, as was the normal procedure when we arrived, but it wasn't there. 'Who the hell's moved the pump?' he shouted.

Just then Deafy walked over, his shoulders shaking as he giggling his head off. He declared that it was on the other side of his new workshop. Fred looked down at him and remarked: 'It looks like a fucking garage to me.'

Deafy was really excited about it. 'Come and have a look inside, Fred,' he said. We agreed to see it just to get some peace. Deafy was almost jumping up and down with excitement, telling us it had a pit with lights, new workbenches, stores with a lock and key to keep us thieving bastards out. It also had a mess room with a plaque bearing the words 'Fitters only – riff raff keep out.'

On entering, we just stood there spellbound. It was absolutely fantastic. It even had an overhead crane. Deafy was like a little boy with a new toy. He went as far as showing us the windows, telling us they opened as well. Banksy had done Deafy and the company proud. Fred put his hand on Deafy's shoulder, and told him he approved of his new workshop. That kept him happy for a while.

I stood there thinking to myself about how the fitters had worked in the old shed before. God only knows how, but they'd managed all right. It had no windows, and when using the pit, you had to bail the water out with a bucket first.

Then there was the brand-new pump. I could not believe it – this one actually ran on electricity. I put the nozzle in, pulled the handle down and that was it. No more turning that bloody handle.

Banksy was also going to have the yard and the roads which led into it concreted properly. Fred shouted to Deafy, 'The money we make for the old bastard, it's a wonder he hasn't had the shit house gold-plated as well.'

Fred went to the office to collect our wages and sort out the expenses. He told me, 'You're bound to hear a lot of yelling, because I'm telling Banksy about the fortnight's holiday I want.'

Deafy asked me to park the lorry, so I knocked the pilot switch down and started the engine. I hadn't backed the lorry and trailer into such a confined space before, so I began to shunt. Deafy wasn't much help because all I kept hearing him shout was 'Mind the pit! I don't want you putting your wheels down there.'

In the end I had to tell him to shut up. After a long struggle I managed it, although it was crooked.

'You did well there Shay,' said Deafy. 'There's just one problem. It's arse about face. I want to work on the engine, not the trailer, you silly bastard. The bench is at the back. Shay, use your common sense. Drive it out, then drive back in, so that the cab is facing the bench. Okay?'

So I did just what he said. But as I jumped down Deafy shouted, 'Why did you bring the trailer in? I told you I'm working on the engine. Back it out again and don't go down the pit.'

By now I was so furious and frustrated that it was a miracle that I kept my hands off him. The perspiration was just pouring off me. I heard him mutter, 'Boys today have no idea at all. I thought it was me who was deaf. Now I'm not so sure.'

That made matters worse. I backed the wagon and trailer out, telling myself to keep calm. I jumped out of the cab, unhooked the trailer, then drove the eight-wheeler into the garage again. The steering was so heavy that my arms ached. As I looked out of the windscreen, I saw Deafy unscrewing the radiator cap. Poking my head out of the window, I heard him say, 'That looks all right.'

'What do you mean, it's all right? Is that all you're going to do?' I shouted at him.

'Don't get aerated, Shay. Just back her out and connect onto the trailer. I want to check the rear lights now.'

'You go and fuck yourself. I've had enough,' I said.

'Do as you're told, otherwise I'm seeing Mr Banks,' Deafy said angrily. Little did I know, he was trying desperately not to laugh.

I hit the starter button, knocked the ratchet handbrake off, leaned out of the cab and started to back out. Deafy kept shouting out, 'Watch the pit! Watch the pit!'

I stopped the lorry and in pure frustration I yelled as loud as I could, 'Fuck the pit!' I then proceeded to reverse as I called, 'Deafy, see me back and connect up onto the trailer.'

'I'll do it under sufferance,' he said. 'Fancy getting all grouchy with me, you young whippersnapper.'

Banksy and Fred walked out of the office and strolled towards us. As they approached, Deafy shouted up to me, 'Shay, back the lorry and trailer into the garage.'

'Right oh,' I replied, grimacing. Banksy and Fred stepped back. I started her up yet again for the umpteenth time that day. Taking a wide berth, I positioned the lorry and trailer so that both were in line, then started to reverse. I was very nervous at the thought of backing the lorry and trailer in front of the others.

Banksy yelled something. I tried ignoring his shrill voice, as I guessed what was coming: 'Look out for the pit!' Deafy, to my amazement, told him to leave me alone and that I knew what I was doing. It was more luck than judgement, but it went in like a dream.

As I climbed down, I heard Deafy talking to Banksy. 'It's a pity you don't employ more drivers like Shay. He's a good lad.' Deafy winked at me and nodded his head with approval. It wasn't until that moment that I understood that this had all been his way of getting me used to manoeuvring the lorry and trailer. In reality Deafy was a great bloke.

Banksy said he'd like a word with me. 'As you probably know, Fred's taking a fortnight's leave, so I was wondering if you'd like two weeks off as well?'

'That's fine by me. Thanks,' I answered.

Banksy had two new six-wheel AECs coming home some time the following week. They'd already been painted and sign-written, so he was just waiting for the licensing authorities to get in touch. Reg was having one, so he could drive Fred's lorry for the first week, and Sharpy would drive Reg's.

'I'm not sure about Sharpy driving the eight-wheeler though,' said Banksy. 'That Atkinson is my pride and joy. The last week the lorry and trailer will be in the garage, so Deafy can get it ship-shape ready for when Fred comes back.'

Then Banksy added, 'Fred tells me you want a driving licence. Just send off to Maidstone for a provisional one and leave the rest to me.'

I was delighted. 'Thanks. I appreciate that.' Looking at me, he said it was a pleasant change to find someone who appreciated him, as many of the ungrateful bastards who worked for him didn't. He walked off, and for that split second I actually felt sorry for him.

Fred put his hand on my shoulder and remarked, 'You see that man. He's a better actor than Clark Gable.' That soon put the smile back on my face.

Fred handed me my wages and expenses, which came to more than I was expecting. As Fred had sprung the holiday on Banksy like that, we'd get the rest of our pay when we came back. 'Come on,' Fred said, 'let's get our night-out cases and get out of here.'

As we walked to the top of the lane, I asked Fred what he'd be doing for the next fortnight. 'Or shouldn't I ask?'

'Cheeky sod! Well, I'm going to treat myself to a well-deserved new suit, have an early night, then catch the early rattler up to London. And then, young man, it's Lancashire here I come.'

As we stood and shook hands, I thanked him for seeing Banksy about a licence for me. He said it was no problem, but to be sure to pass. Fred also made a point of thanking me, saying: 'If it wasn't for you, Lancs and I wouldn't be together.'

As we went our separate ways I whispered under my breath, 'Be happy, Fred and Joyce.'

As I was waiting for the number 480 bus to

Gravesend, I started to think about everything that had happened lately. Fred had met Lancs, I'd learned to drive, Banksy had had a new garage built and more new lorries were arriving.

There certainly had been quite a few big changes.

Chapter 7

The Two Vagabonds from the South

I SOON came to the end of my first week's holiday. It had been a nice break and a good rest for me. Sitting in the front room of my house, I heard a lorry pull up outside, and I immediately knew it was Fred's by the sound of the engine. I opened the door before Reg had a chance to knock.

'I've got some good news for you, Shay. Banksy is going to take you for a test on Tuesday afternoon. Bring your provisional licence with you if it's come by now. But don't worry if it hasn't – Banksy will get it on a pink form, then pin it to your provisional licence afterwards.'

'That's a stroke of luck, Reg. My licence came in the post this morning.' But I was puzzled. 'It's a bit sudden, isn't it? How can Banksy take me for a test so soon?'

'Oh, Banksy's very well known in the ministry and civil defence. He's known some of them from as far back as the war and he knows who to keep in touch with and who not. Also, he's high-ranking in the Free-masons, which helps – but keep that under your hat.

'So, young Bill, he would like you to be at the yard by nine on Tuesday morning.'

'I can't believe it, Reg!' I replied, rubbing my hands together.

Reg winked at me. 'Between you, me and the gatepost, he's had a lot of pressure from me, Deafy, Sharpy and Fred. But I know for a fact that deep down Banksy likes you.'

'Reg, you've really made my day. Thanks a lot.'

'Well, I'll be on my way. I'm off to the Merry Chest now. Hope to see you some time. I know the missus would like to see you again, so don't forget us.'

'I won't. Thanks again, Reg.' As he pulled away, it seemed strange not seeing a trailer on the back of the truck.

I arrived at the yard at about eight on Tuesday morning. Deafy was in the garage, so I sat and chatted with him for a while. I told him that I was really apprehensive about the test, but he soon put my mind at ease, settled my nerves and wished me good luck.

The governor arrived at eight-thirty: 'Morning, Shay, you're nice and early. I won't keep you a minute, I'm just popping into the office to check there's nothing urgent in the post.'

When the old man eventually came out of his office, we walked over to where his car was parked. He immediately jumped into the passenger side, saying to me, 'You drive, Shay.' Before getting in I noticed there were no L-plates on the car. When I mentioned this I was promptly told by Banksy that he couldn't hear a thing and to stop my muttering and get going.

I had just reached the top of the lane when he leaned over and whispered in my ear, catching me unawares: 'Did Sharpy fuck my car up?'

'No, Governor', I replied.

'Lying little toe-rag!' he muttered. 'I'll find out one day, you mark my words.' I refrained from answering further. Much to my relief, he didn't mention it again.

As I drove through Northfleet, he told me to turn right when I passed a church further up on the left. I recognised Pepper Hill immediately; I had been there before with Fred. It led to the A2. He told me to turn right when I reached the bottom of the road, then to make my way towards the Merry Chest and pull in. 'Bloody hell,' I thought, 'he's going to buy us tea!'

As I pulled into the café car park, Banksy suddenly turned to me and said, 'Shay, stop pussyfooting! Drive with more self-esteem and confidence. Don't worry your head about what I'm thinking. You have to be in charge of the car, not the other way around. Don't hold the steering wheel too tight, just relax, otherwise you look like a pregnant fairy.' That made me chuckle.

Banksy continued: 'Right, let's get going. Don't mess about this time, put your foot down, lad.' I did just that, throwing him back in his seat as I went, very nearly taking off. The wheels spun round so fast I burned rubber. It gave me a real buzz. The governor went quite pale. He was probably wishing that he'd never opened his mouth. But when he got over the initial shock he said, 'That's it, Shay, enjoy your driving.'

He told me we'd make our way to the army barracks at Woolwich Arsenal now. Thirty-five minutes later I was pulling in at the main gate. Mr Banks told the guardsman that we had an appointment with Mr Jackson, who was the civil administrator of army communications. The guardsman checked his list, then directed us to his office.

I stayed in the car and watched as Banksy climbed up the wooden stairs and walked across a wooden platform into the main building. The wait seemed never-ending.

Stupid thoughts started going through my head about what Reg had told me earlier about Freemasons, I'd have found it quite amusing to watch the pair of them standing with one bared nipple and a trouser leg rolled up. I couldn't see myself ever joining anything like that.

At that precise moment a fifteen hundredweight truck pulled up. A soldier jumped down and tapped on my window, making me jump out of my skin.

'Sorry I startled you. But are you William Hedley?'

'Yes.'

'Good. Mr Jackson's told me all about you. Let's go to the MT section.'

We drove to an extensive site with army trucks parked everywhere. Once we were inside the office the soldier introduced me to a Captain Taylor. We shook hands. He picked up his cane, tucked it under his arm, and said, 'Let's get on with the work in hand, old boy.'

We stood outside for a while looking at the different vehicles.

'Which one would you prefer to drive, Mr Hedley?'

I pointed to a six-wheel Leyland Hippo.

'That's settled. We'll take out one of those.'

I jumped up quickly behind the steering wheel and sat there familiarising myself with everything while the captain climbed up into the passenger side. 'Right,' he said. 'Let's see what you can do, shall we?'

I did the necessary and gently pulled away from the MT Section. I asked Captain Taylor where I had to go and he told me to follow the road in front. After about ten minutes I drove past the MT Section again. The camp was so huge. He suddenly shouted at me to brake and the Hippo screamed and shuddered to a halt.

'Good, good!' he shouted across to me. 'Now you can take me back to MT.'

I pulled up in the square alongside the other trucks. 'Well, I must say you certainly know how to handle this truck,' said the captain. 'I thought the hand-brake might have confused you. All the new recruits complain because they think it's broken.'

'I must confess, sir, that I have driven – if you can call it that – around Mr Banks's yard once or twice and shunted a few times.'

'Oh, I see,' he replied, then went on: 'If you are driving along a road and you see a sign with a letter "P" on it, what does it mean'?

'Parking,' I answered.

'Well done, old boy! We'll make our way to my office now.' When we arrived back he asked the clerk for a pink form, on which he wrote 'Passed A, B and C', then pinned it onto my provisional licence. I thanked him, then the driver of the 15 cwt took me back to Mr Jackson's office.

As we approached, Banksy and Jackson were standing on the wooden platform waiting for me. Mr Jackson and I shook hands. The motor transport officer had been very pleased with my performance, stating that I was a natural behind the wheel. As we said goodbye, I grinned to myself because Mr Jackson shouted 'Cheerio, Sydney' to Banksy. It sounded so strange to me.

I thanked Mr Banks for getting me a licence. He did mention the fact (which I already knew) that because I wasn't yet twenty-one, I still wasn't qualified to drive a vehicle that was over three tons empty, but as soon as I became of age he would find me a decent lorry. He also said that if he took me off Fred's lorry and gave him another trailer boy, it would cause World War Three, and he didn't want that.

Fred had made plans to be away from home for almost a complete year. That way he could do his job and see Joyce from time to time. He knew that I could work with him because I was happy to work away, unlike many other young men. Fred did stress, though, that whenever we collected a load, the rates had to be good otherwise Banksy would tell us to come back home.

It was good news for Fred and me that road transport was up-to-date and gave good service. British Rail were set in their ways and didn't like change. Over the years they had had the monopoly of transport, but now many of their customers were using heavy haulage companies instead. The work-load was far more in those days, and lorries were scarce.

Banksy offered to take me home and as we drove along we chatted. 'I don't know whether I mentioned it, Shay, but I'm scrapping the Vulcan and the next smallest one is the four-wheel ERF, but that's way over four tons. I meant what I said: if I had a lorry that only weighed three tons empty, there wouldn't be a problem. You could drive it. But never mind, you'll get old soon enough.'

'Not to worry, Mr Banks. I love working with Fred anyway.'

When I got home my dad couldn't believe it when I told him I'd passed my driving test in a six-wheel Leyland Hippo. He thought it very strange that I'd received a provisional licence one day and passed the next. My old Dad wasn't silly. I heard him mutter under his breath, 'There must have been some fiddling going on somewhere.' Naturally he was suspicious because I was under twenty-one. I thought proudly: the next time I'm home I'll be driving one of my own.

Regrettably my second week's holiday was nearly over. I was racking my brains trying to think of somewhere to go when I remembered Rosie.

The following morning, as soon as I boarded the Canterbury train I began to daydream about her. Just remembering the last time I was with Rosie made me feel really randy. As the train sped along the track, it seemed to be saying, 'She won't be in, she won't be in.' My brain was trying to turn it round and I started willing it to say, 'She will be in.'

After the train eventually pulled in at Canterbury West station, I made my way towards St Peter's Street where Rosie lived. As I stood outside her house, I began to get cold feet, and was just about to change my mind when the door opened. She was so astonished to see me standing outside that she nearly dropped the letter she was holding.

'Oh, Shay!' she cried out. 'You frightened the life out of me.'

'Sorry about that, Rosie. Anyway, how are you? Long time no see.'

'I'm fine thanks,' she answered with a twinkle in her eye.

She told me that she was just about to post a letter, but it could wait until later. 'Come on in.' She put on the kettle. As we sat talking and drinking our tea, she told me she'd given Sharpy the elbow, saying he was too arrogant and bombastic. She liked a man to be a man but he was too much.

I must admit my mind wasn't really on what she was telling me. It was more on the curves of her body.

'What brings you to Canterbury, Shay?' With slight hesitation I answered, 'You, of course.'

'Ooh, that's nice!' she giggled. With no more ado she slid her chair closer to mine, and as we chatted she put her hand on my knee, stroking it. She murmured to me, 'I expect I've told you this before, Shay, but your body is so young and lithe. I want to undress you and bathe you.' I didn't object. She guided me upstairs to the bathroom and filled the bath.

One thing led to another in the natural course of events. Afterwards we lay exhausted and finally slept.

I awakened to the sound of the kettle boiling in the kitchen, so I clambered out of bed feeling a little the worse for wear. What a day! My legs felt like jelly. In the kitchen, Rosie asked me if I fancied going down the local for a drink, as she didn't get the opportunity very often. She didn't think it right to drink alone.

After we returned from the local, she asked me to stay but I made an excuse, saying I was going up north and had to leave early in the morning. She was genuinely sorry to see me go, but I promised her I would come back to see her some time in the future. As we kissed and said our goodbyes her hands started wandering, so I pushed her gently away and made my exit.

Monday morning I arrived at nine, refreshed and ready for work. Fred was already there, looking round the lorry which was gleaming and smart after Deafy had polished it. He had even repainted the bottom half of the lorry and trailer. I was pleased to see Fred. Before I could ask about his holiday, he said, 'I'll tell you all about it later.'

Banksy had given us a load out of Associated Portland Cement, Greenhithe. As we travelled through

Stone, Fred suddenly piped up, 'It isn't right, is it, Shay? Just come back off bloody holiday and get lumbered with a full load of cement, and I bet it has to be delivered somewhere round the back streets of London.'

We made our way to the transport office. 'Well, lads you've got a choice! They're both full loads. Coventry or London?'

'Coventry!' we answered together.

After the office clerk had handed Fred the delivery note we headed for the loading bay. Fred backed both the trailer and eight-wheeler under the conveyor belts: that way they could be loaded at the same time. We stood and watched as the two fellers loaded us. They made it look so easy. When there wasn't much space left, one of them pressed a button which made the conveyor go back. That way they didn't have to walk too far when loading. The load stood four bags high; one bag was on the outside with another on top, which pinched the load in towards the middle of the vehicle so it didn't need to be roped so much. They made a good job of loading and we were soon ready.

We hadn't eaten so we had breakfast in their canteen. Two breakfasts and four teas came to the princely sum of one shilling, which was a lot of money in those days for a works canteen. I didn't think the price was too bad but Fred did. He remarked that they could poke their breakfast and tea up their arse in future – a typical driver's comment.

Before we climbed up into the cab we brushed off our overalls with our hands because they were absolutely smothered in cement powder. 'I should have got a job in a flour mill and had done with it,' Fred commented.

As we sat in the cab, Fred made the log sheet out for the day. Then he pushed the starter button and once again we were on our way, this time to Coventry via London. 'I expect we'll be calling in at Lancashire afterwards, Fred?' I remarked with a grin.

'You know what, Shay, the older you get the feistier you become.'

We headed towards Stone crossing, then turned off onto the A2. The road was clear for a change, so Fred quickly pulled across and followed the signs to London. It was hard work for the old girl, pulling all that dead weight.

It certainly was a long haul to Blackheath hill. As Fred drove downhill, I automatically took up three notches on the hand-brake. I knew it would be split-second timing either way, because if the lights changed to red, I had to react quickly and wind the bastard up, and if they changed to green, I'd have to knock the hand-brake off. Then Fred automatically pushed the accelerator pedal down as hard as he was able. If our speed was too high coming down and the lights suddenly changed to red, Fred would never stop in time. It was purely the driver's timing and you couldn't afford to make a mistake. It was because of the weight and vacuum brakes.

The Gardner engine was vibrating the shop windows and doors as we drove along the Old Kent Road. Fred explained to me that the thrust from the exhaust was echoing in the confined space. Heads turned in our direction as we passed.

Since leaving Blackheath, Fred hadn't managed to get out of third gear. It was really slow going. Fred decided to drive through Bank in the city on our way to Barnet.

'How did this place get its name, Fred?' I asked him.

'It's because the Bank of England is situated here. There's certainly a lot of history in London, it's an education in itself.'

Fifteen minutes later Fred pointed out Holloway prison to me, which was for women only. As he drove past I took a really good look, not having seen it before.

Still heading for Barnet, Fred turned left into the famous Archway of London. Once on the other side he crossed over the A1, then onto the A6. Before long he was driving through St Albans. We must have been travelling on the A5 for roughly three miles when it started to drizzle, and as it hadn't rained for a long while the road was very greasy.

The car in front of us suddenly stopped dead. 'Fucking hell!' Fred cursed, braking hard and slithering to a halt behind him, missing him by about a quarter of an inch. I noticed that two cars travelling in the opposite direction were rubber-necking. The second car was so busy looking across at us that he hit the car in front of him up the jacksy. Crunch!

Then suddenly a five-ton petrol Austin lorry slid to a halt behind him, but the fully loaded AEC four-wheel petrol travelling behind the Austin couldn't stop and smashed straight into the back of the lorry. The AEC pushed the lorry with such force that it crashed into the back of the car that was rubber-necking and had caused the accident in the first place. Now the car was scrunched and looked like Emu at the front. It was in a right old state.

Fred jumped down from our cab, shouting to one of the drivers to ring the emergency services pronto. When he reached the driver of the four-wheel AEC, he saw that the cab of his lorry was completely smashed and tried to help, but another driver told him that he might do more harm than good if he moved the driver.

I could hear the poor driver screaming out with pain. It was horrific. The steering wheel had almost embedded itself in his chest, and his left leg was trapped. Fred and I tried to calm the poor sod until the ambulance arrived.

It wasn't long before one of Allison's of Scotland drivers joined us. He asked Fred if there was anything he could do. 'You can move those two cars out of the way,' said Fred, 'and ask the driver of the Austin lorry to move up at least five foot. If he'll do that we can attach a chain around the back of his lorry and put it round the steering post and pull the bastard back.'

By this time quite a few people had gathered round. Fred had taken the initiative to organise things. The driver of the first car had managed to park up on the grass verge; the rear of his car was completely crumpled. Fred told the driver of the second vehicle, who'd caused the accident in the first place, to move up out of the way.

He promptly told Fred he couldn't as it might damage his engine, because the radiator had got stuck in the fan.

Fred grasped him by the lapels of his jacket, lifted him off the ground, and said, 'Mister, if you don't move this car it will be you they'll take away in the ambulance. So move it!' While the man's legs were dangling in mid-air, his little booties were knocking together. He looked to me just like Pinocchio. I felt sorry for him really, because Fred was in such a temper I thought he might kill him.

At that moment the driver from Allison's, whose name was Jock, jumped in the car and started her up. Bang! It sounded just like a gun going off. We realised that the fan blades had completely disintegrated, and the exhaust had also been damaged. But Jock did manage to move it up along the road out of the way. Water and oil were pouring out from underneath the vehicle.

The five-ton Austin started to pull away from the crashed lorry. Fred waved him on, telling him to take it very slowly. Once the lorry was out of the way, Fred signalled to him to stop. I retrieved the horse chains from the toolbox, taking them to the scene of the accident. Fred took them from me and tied one end to the back of the lorry. Jock tied the other end onto the steering post.

Fred told the other driver to get back into his lorry and get ready, and to make sure and pull away very steady and keep him in sight at all times. Fred first checked on the injured man to make sure he was comfortable. Then he beckoned to the driver to pull away slowly. The steering column moved forward by at least a foot. Fred immediately signalled again for the driver to stop.

At that moment the ambulance finally arrived, so Jock, Fred and I gave the ambulance men a hand by putting one of the chains round the clutch bar and pulling back gently to free the injured man. He was then taken away in the ambulance.

When I arrived back at the scene after putting the chains back into the toolbox, Fred was questioning the driver who had stopped dead in front of us. I overheard him tell Fred that his engine had completely seized up, locking the rear wheels. Fred asked some drivers who were standing around to bump the car onto the side of the road, which they did.

Not long after that the police arrived. Fred had written some of the particulars on a piece of paper which he handed to the officer in charge. Then he turned to me, saying, 'Come on, Shay, let's get cracking. We've got work to do.' We waved to the Allison driver and were once more on our way.

By now I knew every inch of Watling Street on the A5. I had even got to know a few of the night trunkers. I was becoming quite an old hand.

That night we would be stopping at Bob's transport café in Coventry, and from there we'd pop into the Dun Cow for a few jugs. Fred reckoned we might even be able to flog some bags of cement. He was always on the lookout for some scam!

It was a long haul to Coventry, and the road was a little the worse for wear. As the front wheels hit the grooves and holes in the tarmac, the steering wheel shook unsteadily, making it difficult for Fred to control the lorry. We eventually pulled in at Bob's for the night.

Since we were ravenous we decided to have a meal before a wash and brush up. Just as we sat down to eat, one of the lads from Birmingham approached us: 'Would either of you be interested in buying a settee? Latest design, brand new.'

Fred told him we had a full load of cement on board so there wasn't room for it even if we wanted one. Also, we wouldn't be home for at least another week.

As the guy walked away, I mentioned to Fred that I wouldn't have minded one for my old mum. His reply was: 'Shay, if you think I'm going to carry a poxy settee around for days on end, you're very much mistaken.'

We just happened to be there on one of those nights when a lot of goods were being sold on the cheap. If your Aunt Bessie had been there, she would have been auctioned off as well! I did laugh when one of the drivers asked if we wanted to buy a cooker, the latest on the market. The dealing reminded me of Petticoat Lane, the only difference being that the goods had come straight out of the factory.

One of the Dunlop drivers was selling Wellington boots. As they were going cheap, Fred and I bought a pair each. We heard him say that the dockyard workers had lifted a lot of them. There were wellies of all sizes everywhere.

The boots were selling like hot cakes. I thought to myself: by the time he gets to his destination all he'll have is empty boxes. I could just picture the customers' faces when they finally opened these cartons from Great Britain, full of nothingness.

As we walked through the door of the Dun Cow the landlord recognised Fred immediately. Will you tickle the old ivories for us tonight and brighten things up?' he asked, adding that the beers would be on him all evening.

Fred said he'd take him up on it, then leaned towards him and whispered, 'Do you know of anyone who wants to buy some cement on the cheap?'

'How many bags have you got?' asked the landlord.

'I won't know that until we leave the site tomorrow,' Fred replied.

'One of my regulars is in the building game. When he comes in tonight I'll ask him.'

'Thanks a lot. I'd appreciate that.' Fred then seated himself at the piano. He glanced over to me, winked, then proceeded to play 'We're in the Money'.

As I supped my beer, more drivers walked in. During their conversation I heard one say that they did more miles and burned more diesel sitting in the pub than they actually drove on the road. Some had been on the road for days at a time.

Old yarns were being told. One driver said, 'What do you think! I picked up this woman and as I was driving along the road, I discovered that she was a witch.'

'Why did you think that?' we chimed in, on cue.

'Because she turned me into a lay-by!'

Then another piped up, 'What do you think of my watch, lads?' As we took a closer look, we saw it had a variety of dials on the face. He then told us: 'Press this button on the side and it gives the time in nearly every country in the world.'

After about ten minutes of discussing his watch, he suddenly shouted, 'Fuck it!'

One driver asked, 'What's wrong with you?'

Peering down at his watch, the man said, 'Guess what? It's raining in London.'

They shouted back, 'You're an idiot, do you know that?' We all laughed.

The evening passed very quickly and in no time at all Fred was playing his old signature tune, 'There's an Old Mill by the Stream'. As usual I went around with my collection pot, and then it was time for home.

In our digs at the top of the stairs there was an enormous dormitory which held at least forty beds. The place was spotless and the bedding clean. I had a wash, climbed into bed, and in a short while I fell asleep, dreaming of nestling my head between Rosie's tits. They were enough to make anyone deaf, let alone sleep.

We arose at six the following morning, tucked into a hearty breakfast, and left Bob's café at just after seven. Our delivery was in Baginton near Coventry airport. Just before we arrived at Baginton, Fred pulled up in a nearby lane, saying he would drop the trailer there. When I asked why, he said with a wink, 'Use your loaf. There's at least four tons of cement on, so we'll sell it to that fellow the café owner told us about and split the difference. Okay?'

When we arrived at the site the charge-hand told us to drive down to where all the other cement bags were piled high and remove our lorry sheets and ropes, saying he would send the gang down. When they pulled up Fred and I gave them a hand with the hundredweight bags of cement. We unloaded fourteen tons of the stuff. A young lad in his twenties signed our ticket.

As Fred pulled away from the site I felt a bit guilty, but Fred just rubbed his hands together and started singing, 'We're in the Money'. I soon forgot my guilt and sang along with him.

The builder we were dropping the rest off to lived in Brandon, which was ten miles from the Baginton site. The weight of the trailer pushed hard onto the back of our lorry. Fred drove really carefully as he didn't want to have an accident, especially with the dodgy cargo on board. Eventually we found the building site.

The site owner asked, 'How many bags have you got there?'

'Eighty,' Fred told him. The poor fellow nearly collapsed with shock, but agreed to take them all. He laid scaffold boards down for us. Fred jumped up on the trailer and handed the bags down to the pair of us. When we'd off-loaded, Fred asked me to tie the sheets down onto the trailer, then turn the lorry round while he discussed the money side with the building site owner.

As soon as Fred climbed up inside the cab he handed me three pound notes. I thanked him, saying it would pay for a good many breakfasts and that I might even be able to save some.

Our next port of call was the Coventry Climax factory, to ask if there were any deliveries going for the north. However, when Fred appeared from the office, he exclaimed, 'Isn't it bloody marvellous, all the loads are for the south! But unfortunately for you, Shay, I've managed to get one for Fleetwood.'

'You make me laugh, Fred,' I said. 'Everyone tries to get a load nearer home. But not you, you're the opposite from everyone else. It must be love.'

Luck was running with us, because the stock transfer that was going to another depot was packed in large crates, which made it easier. An overhead crane loaded us, but the best part was we did not have to sheet, just rope up using double dollies.

We pulled out of Coventry at midday, Fred driving through the town centre. The roads were very narrow and when Fred had to turn left, the back wheels of the trailer unfortunately ran over the kerb. It couldn't be helped.

I hated Fred driving through towns as he took great delight in making me do hand signals. If it was raining hard the sleeve of my coat got soaked. I recall one occasion when I had upset Fred. He made me drop the window and told me to give a left-hand signal. His speed at that time was about 35 miles an hour. I even remember the road, it was the A5.

After a while he told me to put my hand back in and said there was no point in me getting arsey as I would never get one over on him. When I look back now it makes me laugh, but at the time I can remember only too clearly muttering, 'Bastard!'

We turned off onto the A444, heading towards Nuneaton, which was a lovely town and the roads were much wider too. Fred hadn't been heading north for long on the A5 when he turned right onto the A51 towards Lichfield.

As we stopped for a tea on the outskirts of Stafford, I asked Fred where we were staying that night. 'Leyland,' he said. 'Actually, I know of some digs. The landlady is great, and looks after you well. She'll even give you an early morning call if you want one and it's not that far from where Lancs lives, either.'

'I thought as much,' I answered.

It was only the large transport companies like the one we worked for that had eight-wheelers. They were very well made. Their weight when empty was approximately seven tons, and when you looked at the eight-wheeler and saw the size of the spring hangars and thickness of the chassis you could see that they were over-engineered. But they were the best vehicles of their day.

The price of these vehicles was exorbitant; only the better-off companies could afford them. Looking back, I think our lorry looked huge, especially with the trailer on the back, and particularly when it was parked alongside a four-wheel lorry.

I seated myself comfortably into the passenger seat, as I knew it was going to be a long haul to Leyland. Very soon we would be travelling along the A6, known to be one of the worst arterial roads in the country. But the council didn't seem interested in repairing it. Our load was square. I preferred it like that because neither Fred's nor my vision was impaired.

Fred had driven at least a mile and a half before he managed to get her into fifth gear. From then on he had to pull every trick in the book to keep the momentum up, because if he eased off the accelerator longer than five seconds, it meant changing down another gear. He had to keep the revs up at all times, which was hard to do with a Gardner as it had a low-revving engine.

As soon as he started to climb a hill the Gardner engine would have died on him if the revs weren't kept high, so Fred had to drive, not pussyfoot the Gardner all day and every day. It was definitely a governor's engine not a driver's engine. But what he did like about the Gardner was that it was a reliable piece of machinery and always got him home.

I could feel Fred's enthusiasm as he made her tramp on, but as soon as we came to a downward hill he knocked her into the silent six so she would glide down as if on ice. As I sat looking through the mirror, the trailer was rocking gently. It was great doing forty miles an hour now and again, but as soon as we came to an upward hill again, she slowed right down to thirty-four, the maximum speed. Fred then had to rev up and put the gear lever into fifth, which meant on occasions our speed was fourteen miles an hour over the official speed limit.

I could now see the signposts for Knutsford coming into view. In some places the roads were very narrow and our speed would drop down to 9 mph. When we reached the other side of Manchester the traffic wasn't too bad as most of the commuters were coming in.

Eventually Fred turned off the A6 and drove into Leyland. He found a place to park, which took some doing when it was busy. When he stopped I lifted the engine bonnet and checked the oil. It was low so I topped it up out of the gallon which we always carried on board. As I was filling it, Fred piped up, 'Don't spill any over the engine!'

Looking him straight in the eye, I replied, 'Do you want to do it?'

'Only joking,' he replied, giving me a wink. At times I still didn't know how to take him. Before Fred locked the lorry I checked the tyres and wheel nuts. We picked up our suitcases and walked up the road. As we had made good time we decided to call on Lancs there and then.

I noticed that the bungalows and houses on the other side of the road were becoming more up-market. I was so busy looking that I didn't notice Fred disappear through the gate of one of the bungalows. Then I spotted him peeping at me, grinning all over his face like a Cheshire cat. When I walked up to him he had the audacity to say, 'What kept you? I've been here for ages.' I could never stay mad at him for long. He was always fooling about.

Joyce wasn't in so Fred put a brief note through her letterbox, telling her he was in the area and would see her tomorrow, all being well. Our digs were only a short distance away. Perhaps it was a good thing she wasn't in, as we must have looked like a pair of vagabonds standing there on her porch.

Fred was right about our digs. It was excellent accommodation. We had just finished our meal when the landlady appeared and told Fred that a young woman would like to see us tonight if we hadn't made other arrangements. We headed back towards the bungalow, this time dressed in smart clothes. We looked very presentable indeed.

'You'll want to be on your own, Fred, so I'll make myself scarce.'

'Don't be daft, Shay. You don't think I'd leave you to your own devices in a strange town, do you? Anyway, she'll want to see you as well.'

I protested, but he wouldn't accept arguments. 'Shay, you're coming and that's that. Anyway, if I leave you on your own too much you'll get fed up and want to go home and I can't have that. You and I, young man, make a good team.' Fred smiled, then whispered, 'You can creep off later on if you like.'

Fred rapped on the door of Joyce's bungalow. The door opened and she welcomed us in, saying she'd seen our lorry and trailer parked up and that was why she'd come to the digs. She hadn't known then about Fred's note.

The wireless was playing quietly in the background as we talked. When Joyce brought the tea in, the bone china cups, milk jug and sugar bowl looked so delicate on the tray.

After a while I stood up and made my excuses to leave so they could spend what was left of the evening together. Lancs walked me to the door then, much to my surprise, kissed me on the cheek and said, 'Thanks, Shay, you're very thoughtful.' I smiled to myself as I walked up the path.

I walked back to where our lorry was parked. I stood there for a while admiring it and reminiscing. To me the old Atkinson had a great deal of character. I loved everything about that old lorry. Then I strolled along to the pub for a pint, chatted to a few of the locals then around eleven made my way back to the digs.

The next thing I knew Fred was tapping me on the shoulder: 'Wakey, wakey, rise and shine!'

'What time is it?' I asked sleepily. I sat on the edge of the bed for a while, gathering my thoughts. 'Come on, Shay, get your knickers on. Time waits for no man, as they say. The landlady was cooking breakfast in the kitchen, the delicious smell wafting up the stairs.

As we sat there tucking into bacon, eggs, fried bread, sausage and mushrooms, Fred asked me if I had my driving licence with me, which I did. 'Let's have a look at it.' He scrutinised it for what appeared to be a long time, then remarked with a broad grin on his face: 'Somewhere along the line, Shay, there's been some fiddling going on here.'

'Is that so?' I replied, trying desperately to keep a straight face.

After leaving the digs we made our way back to old Betsy. As always, we checked to see everything was in order. Just as I was about to jump up into the cab Fred shouted, 'Shay, you take the starboard and I'll sit on the port side for a change. You can drive.'

I climbed up behind the wheel and immediately the adrenaline started flowing. I started the engine – it sounded like music to my ears. I gradually let the clutch up as my legs trembled slightly with excitement. Suddenly I was aware of Fred's voice saying, 'Turn left here, Shay, and make sure you pull right over to the crown of the road. Lug as hard as you can on the steering.'

Fred encouraged me as I drove by saying, 'Well done, Shay, well done.' At the grand old age of twenty I was driving an eight-wheeler and trailer along on the Dartboard and sitting beside me was one of the finest transport drivers in the country. Feeling very proud of myself, I tramped north.

I observed all the signposts that read Blackpool as I made my way to Fleetwood. As I was negotiating a very steep hill – in drivers' language, 'a long drag' – I changed from second down to first gear. As I did so I caught one of the cogs along the way which made a horrendous noise through the gearbox.

Fred didn't say a word, but the look he gave me was enough. I said sorry, but Fred just told me to be more careful and not to rush it.

Then I had to change up again – 'here we go, it's shit or bust,' I thought because for me this was the hardest thing to do. I slipped the gear lever forward into neutral and just as the lorry was about to stop I put her into second and let the clutch up as quickly as possible, giving it full throttle. It went like a dream.

I looked across at Fred who nodded his head in approval but ruined it by saying, 'Don't get too cocky now.'

'Bastard!' I thought as, much to my relief, the road began to straighten out again.

One advantage I had was that over the years I had been watching Fred intently and had got to know the sound of the engine and gears. I realised immediately if there was something amiss.

I eventually pulled into a distribution centre where it didn't take long for the fellers to unload the large crates. While they were off-loading, we had a mug of tea and a sandwich in their canteen. Fred made a few telephone calls and managed to get a return load out of Lancaster for Birmingham.

Banksy was always delighted with Fred as he always managed to get good rates for his work. I was thrilled because Fred let me continue driving. I actually got the feel of the vehicle and riding the potholes was great. The lorry had Michelin tyres on the front. Although they were the best, they always looked flat to me. And though the lorry was empty the steering was still extremely heavy.

As the months moved on Fred allowed me to drive more and more. I loved every minute of it: the driving, the smell of the diesel, the sound – it was my life. Sometimes I drove through the night to get back to Leyland, time just flashing by.

One of my finest moments was as we left the Central transport café in Burnley one morning after I'd been driving for about ten months – illegally of course. Fred put his hand on my shoulder and said, 'Shay, I can't teach you any more about road transport. I've taught you all I know and I look upon you now as an excellent transport man.' When he told me that, I was choked with emotion and elated.

In the ten months of being away from home we had, to the best of my knowledge, been back to Dartford at least five times for servicing.

One morning at breakfast in Spennymoor, Fred told me he would ring Banksy to put him in the picture and update him on the work we had been doing. After a short while Fred arrived back at the table. He looked shaken, and had a troubled expression on his face. 'What's wrong?' I asked.

'Banksy's just given me some terrible news. It's poor old Reg. He's had a heart attack and died.'

I thought I'd been hit by a thunderbolt. Reg Maynard! I couldn't believe what I was hearing. 'Yes, Shay, it's true, so we'll make our way back to Kent as quickly as possible.'

Fred took over the driving. He seemed to be flying along the A1. As I sat in the passenger seat I felt the tears burning my eyes. I could remember only too well the words Reg had said to me the first day I met him: 'The first thing on the agenda is a cup of tea.' Then there was the time he knocked the gear lever into neutral on top of Swanscombe cutting, looked across to me and said, 'Look out for the old coppers, young Bill!'

Sitting in the Merry Chest café at Swanscombe, he had almost wet himself with laughter at all the jokes that were being banded about by all the other drivers. I smiled at the memory of Fred Warren asking him if he had any spare barrels on board, and Reg replying; 'If I had, I would have sold them before I came in here, especially with you lot of old vultures.'

I imagined myself back in his garden drinking lemonade with his lovely wife, Elsie, when all she wanted to do was to get as much food into me as possible.

By now the road outside was a blur. I'm sure Fred knew how I felt; that was why he drove.

We had been on the road for about half an hour when Fred decided to stop. He put his hand in his pocket, took out his address book and proceeded to look for a company he could ring for a load to take back to Dartford, as the governor would have asked in no uncertain terms what he was paying us for. Eventually he managed to get a return load from Watt's Brothers in Beverley, Yorkshire for delivery to Poplar in London.

In what seemed like no time at all, Fred pulled into their yard where the governor told us to load out of Cottingham, which was about ten miles away. They loaded one-hundredweight bags of animal food, eighteen tons in all, out of a warehouse. It was hard work – all hand board.

It didn't take Fred and I long to rope and sheet. The journey plus loading had taken about two and a half hours altogether. As we left the yard. I noticed an eight-wheel Thornycroft in the workshop. To get to the other side of town Fred had to drive all the way round the river Humber. Fred told me he would stop in Doncaster and park up, as it was an excellent place and well known in the transport world. Fred was right as always. There were transport digs from one end of the road to the other, which was brilliant.

Fred and I didn't socialise at all with anyone that night. We just sat on our own in a corner of the pub and chatted. 'What we'll do tomorrow, Shay, is get an early start, and by hook or by crook get this load off as quickly as possible. Then I'll drive to the Merry Chest, book the load off the next day, and you, Shay, can go home on a dodgy night out. How's that with you?'

'Thanks, Fred, that'll be fine by me.'

'But understand this, Shay. On paper we've booked off at South Mimms.'

'Gotcha,' I answered.

At six-thirty the following morning, we had the bar back on the Gardner and were thundering down the A1. By now we had both come to terms with Reg's death and were able to talk about him more easily. Ten hours later we pulled in at Poplar. The foreman there went bananas as the time was four-thirty in the afternoon. The couple of blokes who off-loaded us did it grudgingly as they wanted to get off home, which was understandable, of course. They were miserable bastards all the same – typical dock workers.

After leaving there we made our way to Bean, on the A2. Fred parked in the Merry Chest café. I had an easy morning, reading the papers and such like, and met Fred the next day at 2 pm. We then made our way back to the depot.

As we stood in the yard at Dartford, the old man was very quiet. Reg was his boy, although everyone who knew Reg thought he was a great fellow. Banksy didn't even bother to query Fred's expenses when he handed them to him. Fred said to me afterwards, 'If I'd have known the governor was going to be like that I would have booked another fiver.'

The governor informed us that the funeral would take place at Dartford cemetery. We told him we would be there and he mentioned that he would be paying us eight hours for that day, which was reasonable, I thought. Fred thanked him. After that we made our way to Elsie's to offer our condolences.

Walking up the road, Fred commented, 'You know what's bothering me, Shay? How on earth am I going to get in touch with Lancs to let her know what's happened? She'll be expecting us.'

'That's no problem,' I said. 'Ring the digs where we stopped in Leyland. I'm sure the landlady would notify Lancs for you.'

'Good thinking. You can be intelligent at times, Shay.'

'Sarcasm will get you nowhere,' I replied with a grin.

Chapter 8

The End of an Era

THE day of the funeral was bitterly cold although the sun shone brightly. The church was packed with drivers from all over the country, and all the Merry Chest customers were there. I could see from Elsie's face that she was astonished to see so many come to pay their last respects to Reg.

After the vicar's sermon, a well-dressed gentleman rose from his seat to say a few words about Reg. It was very moving, most of the people there shed a tear. At first I didn't realise who he was, then it dawned on me: Deafy. I didn't recognise him because usually I saw him in his overalls, smothered in grease, no teeth, and wearing National Health glasses. Now he looked a completely different person in his black suit and tie, white shirt and black, highly polished shoes. His teeth were in. He also had different spectacles on. I loved old Deafy.

Later on at the grave-side, I noticed some of the fellows from Truman's Brewery. As Reg's family walked away, the Truman's lads threw a couple of beers into the grave. Reg did like his beer.

The next day it was back to work as usual. We all congregated in Deafy's garage while Banksy rang round to find us some work. Deafy came hobbling over. 'I don't want you riffraff in there. Piss off the lot of you,' he shouted, grinning all over his face.

Sharpy yelled back, 'Don't you start, otherwise we'll put you down a hole and bury you if you're not careful. So watch it.'

Up went Deafy's arms, fists clenched in temper: 'Come on then, Sharpy. If you want a fight, put 'em up – that's if you're man enough, of course.' I had to laugh at those two. They were always at each other's throats, but there was never any real violence.

We all sat down and drank our tea. Glancing out of the window I could see Reg's old six-wheel ERF parked up. I had a sudden urge to sit behind the wheel, which I did later on, because I knew it would be scrapped soon. Gradually the drivers were being called upon to do different jobs until just Fred and I were left sitting there. 'Shay, I'm going to chuck it all in soon,' Fred said suddenly.

'You are joking,' I replied. 'You can't break us up, we're a team. I expect old Banksy will make me redundant.'

'Trust you to think like that. He won't do that,' Fred reassured me. 'But I will say this, Shay, Banksy's days are numbered, because British Road Services will take over all the road transport shortly. At the moment we can get work from all over the country but that will change, you mark my words. Trailer boys won't be needed any more. Most of the work will come through BRS.'

Fred had mentioned a while ago that he was thinking of leaving, but I hadn't expected it quite so soon. 'First Reg, now you. Who'll be next?' I said wistfully. 'What will you do, Fred?'

'Well, Joyce and I will pool our resources and buy our own pub,' he told me. 'We've agreed that we don't want to manage one, and whichever one takes our fancy we'll go for. We've been planning this for ages.'

'Well, with her good looks she'll definitely pull the punters in, that's for sure. And as long as you keep your fingers off the piano and keep serving, you won't lose any custom.'

Fred put his arm around my shoulder. 'Seriously though, Shay, I'll miss you. You've been an excellent trailer boy. You'll always be welcome to visit Lancs and me at any time.' He asked me to keep it quiet for the time being.

The governor came out from his office with the news of a load for us. He told us to load out of Associated Portland Cement at Swanscombe, for delivery at Bexley. Fred looked at him and snapped. 'It's only ten miles away. Haven't you got anything going north? It seems a bloody waste of time to me, roping and sheeting for that short distance. You do take some shit on, Governor.'

Much to my amazement, Banksy answered quite calmly: 'Fred, I can't send you up the road because I haven't got anything to be delivered up that way. And I don't like your tone.'

'What about West Malling aerodrome? Can't you get in touch with one of your left-hand shakers?' insisted Fred.

At that moment the governor went ballistic. He was red in the face with rage. 'What do you mean by that? Left-hand shakers indeed!' I really thought Banksy was going to bust a gut.

All the time Fred was laughing inside because this

was his golden opportunity to throw in the towel. Fred just looked at Banksy, then walked away, saying 'Bollocks.' He grabbed his suitcase from the cab and walked back towards us saying, 'Lick 'em and stick 'em, I'm leaving as of this moment.'

Banksy and I just stood there. I don't know which of us was the more shocked. I didn't know then that Fred had planned it this way. He knew that there wouldn't be any work for up north anyway, especially now that BRS was in force and nationalising all transport companies.

When the governor had quietened down a bit, he told Fred that he was one of the best transport men who had ever worked for him and asked him to think seriously about it and not be too hasty. But what Banksy didn't know was that there was a lovely woman waiting for Fred up north.

Fred told the governor he had thought about it but was still leaving. 'I hope you'll look after Shay, Governor, he's got the makings of a good transport man. As you know, Shay will be twenty-one in a couple of months' time.'

'Shay will be okay. Don't you worry your head about him, Fred,' promised Banksy. Fred handed Banksy his new Leyland address so he could send Fred's cards on to him.

Fred and I turned to one another. 'It's come more quickly than I intended, Shay! But I couldn't waste such an opportunity.'

I didn't blame him in the least. I knew that if I'd been in his position I'd have done the same. I'd never met a person who lived in digs permanently like Fred. Now things were changing and I wished him luck.

Fred went off to tell Deafy, who was dumbstruck. 'I must say this, Deafy, I'll miss you,' said Fred.

'Me too, you old bastard,' Deafy replied as they shook hands. 'Just one thing before you go, Fred. You've done more miles in that old lorry than all the other drivers put together, and you've had the fewest brake linings and repairs. Believe me, it didn't go unnoticed by the governor. Do you realise,' added Deafy, 'I've only got another three months myself before I retire.'

'The way things are going here I'll be the only one left, and the oldest,' I quipped. That made them laugh. Then Fred and I walked out of the garage. We shook hands, and I was a bit choked as I watched him disappear up the lane.

As I stood there in a daze, Banksy rushed out of his office shouting, 'Shay! I'm in the shit so I'll have to take a chance. You're only a couple of months away from your birthday, so can you deliver a load of cement for me? You can either take Reg's lorry or Fred's, the choice is yours. But if you take Fred's you'll have to drop the trailer.' Without hesitation I said I'd take Fred's.

Within minutes I was driving out of the yard. It was still very cold but the sun was shining brightly. I felt confident but a little apprehensive at being on my own. There weren't many drivers my age at that time driving eight-wheelers. It wasn't long before I was pulling in at APCM. I collected my delivery notes and drove under the conveyor belt, then sat in the cab while they loaded me.

It seemed strange sitting there filling out a log sheet because Fred had always done it. I roped and sheeted the load, jumped back inside the cab and drove along until I reached the main road, then I turned right. It took all my strength to turn the steering wheel, and as I let go of the wheel it spun round like a top, which was a new experience for me. It then righted itself.

I started to climb a steep hill, so changed down to a lower gear. As the lorry began to pick up, I tried putting the gear lever into second but the engine stalled, then stopped. I restarted the engine and managed to engage second gear, putting my foot down hard on the foot-brake to stop the lorry from running back. For that split second I had started to panic, then in my head I distinctly heard Fred say: 'Come on, Shay, you can do better than that.'

I put the ratchet hand-brake on hard, put her back into crawler gear and, as I felt the clutch take the strain, I thrust the hand-brake forward and knocked the brake off. She leapt forward a couple of times before I managed to drive away smoothly. 'Thanks, Fred,' I said aloud.

I certainly noticed the difference with fifteen tons of cement on board. This time I kept her in a low gear until I reached the brow of the hill, then I thrust the gear lever straight back into second. This time it engaged like a dream, but my armpits were wet with perspiration.

Just before reaching Stone, I made my way onto the A2, turned right then drove towards the Black Prince. A hundred yards up the road I stopped and asked someone directions to Bourne Road, It was my lucky day because he directed me straight to it, and in no time at all I was unsheeting my load. Within the hour I was driving back out of the site.

I was beginning to feel hungry as well as thirsty, so I carried on to the Elms café on the A2 at Dartford, where I sat down and enjoyed a good, hearty meal. I sat there gazing out of the window at what had once been Fred's lorry. It looked bloody handsome. Once back inside the cab, I felt all warm inside, my hunger pangs gone. I was definitely cab happy and feeling very proud of myself. Time was marching on so I made my way back to the yard.

As I pulled up outside the main office I noticed three other drivers in the yard, so I guessed that old Deafy had spilled the beans about Fred and that I was driving his lorry. As I jumped down from my cab, Banksy greeted me, looking quite concerned. He asked if everything had gone all right. I told him it couldn't have gone better.

'Shay, I'd like a quick word with you in my office,' Banksy said. As I sat down in the office, the governor told me how much he had appreciated what I had done. 'As you are aware, Ruddock dropped me right in it. He can be a funny bastard.' As I sat there, I thought Banksy was the odd-ball, not Fred, and if he had treated Fred fair and square, and paid him the right wages and stopped haggling over expenses, his attitude would have been completely different. I know he would willingly have worked the governor a week's notice.

Banksy interrupted my thoughts. 'How would you like to carry on driving all the time, Shay?'

'I'd love to, Governor,' I replied.

'Everybody would automatically think you were twenty-one, but if anybody did happen to ask, that is what you are. Oh, and by the way, I've got a new driver starting tomorrow.' Banksy asked me if I wanted to take Reg's lorry and work on Truman's beer, or drive Fred's.

'If you take the Atkinson, you'll be away a lot on

journey work. It won't have a trailer, as I'm going to hire it out to Campbell's Wax Company. They'd like it for internal work, and they have their own Ferguson tractor to pull it. That'll be a nice little earner for me. So I'm giving you the choice of lorry.'

As quick as a flash I answered, 'I'll take the Atki, please.' Banksy made me promise I'd look after it.

'Of course I will, Gov.'

'Tomorrow you'll be loading cables, out of Henley's, Gravesend. They'd like you there as soon as possible in the morning and your final destination will be Exeter, so you can take the lorry home tonight. On the way out, see Kay for all your night-out expenses. Don't forget to check that Fred has left the agency card for diesel, and make sure you have daily log sheets.'

Now I felt full of piss and importance and very manly. I started to feel more confident in myself. I asked Kay for the money for four nights out. 'I thought you were only going to Exeter? Will you need as many as that?' I looked her straight in the eye, saying, 'What happens if I get a return load in the opposite direction? Two nights out won't be any good, will it?'

'I'm sorry, Shay, I didn't mean anything by it, but I get told off by Sydney if I pay out too much.'

'Tight bastard,' I muttered, then: 'Whoops, sorry.'

'That's all right,' Kay smiled as she handed me my night-out money.

As I was leaving I said to her: 'You know what, Kay, I dreamed about you last night.'

With a surprised look on her face, she said, 'Did you?'

'No, you wouldn't let me.' That really made her laugh.

It was a lot better filling up at the fuel pump as I didn't have to turn that handle around any more.

As I walked into Deafy's garage, Sharpy was having a crack with Dave Evans and Bill Warnett. Seeing Bill was a surprise. Nobody ever knew where he was. He was definitely a man of mystery, the invisible man – there one minute gone the next. When it was slack and the drivers were working in the yard, you could guarantee Bill would be missing. He was definitely a clever bloke, trained by the lads in the British army. But when the governor was around it was a different kettle of fish. He'd suddenly appear from nowhere.

'We know Fred's gone, Shay. Any other news?' said Bill.

'Well, the governor's employed a new driver and he starts tomorrow. And now I'm twenty-one,' I said, feeling the blush coming to my cheeks, 'he's given me Fred's lorry.'

Sharpy quipped, 'You'll have to have a trailer boy then and that means you'll have one for the trailer and another boy driving a man's lorry.'

'I'll dismiss that remark, Sharpy, as I proved I'm a man with Rosie.' That shut him up. They all agreed that Banksy's was going nowhere, but that I should get some experience under my belt, then look around for a decent company to work for while I was still young. Dave said he was fed up, and thinking of going back to Wales. 'Banksy may treat you like the best thing since sliced bread, but he uses people for his own means. Be warned now, watch the bastard.'

'Point taken, Dave. I'll always bear that in mind, but don't forget, lads, I've had a good teacher in road transport, the best: Fred Ruddock.' I jumped up inside the cab of the Atkinson and as I pulled away I glanced through the mirror, knowing they would be watching me. I was right.

In the cab I found Fred's little black book with the places for return loads, as well as addresses of all the transport digs across the country. He must have left it there by mistake, but I would make good use of it and next time I was in Lancashire, I'd make sure I gave it back to him.

It was impossible for me to explain to anyone how I felt about driving the eight-wheeler. I would be the first to admit, even to myself, that I was cab happy: even after four years I still loved it.

I drove down Bath Street in Gravesend and parked the lorry opposite the Gravesend Ferry. As I walked along I cut through the graveyard, in which stands a beautiful statue of Captain Smith's wife, the famous Red Indian girl, Pocahontas, daughter of a well-known Indian warrior.

As I approached my mum's house, I thought to myself: Mum will be wondering what's going on, with me being home twice in one week. I didn't mention to my parents about Fred leaving the company or that I was driving an eight-wheeler because, knowing Dad, he would not have thought it right, what with me being under-age. Least said the better, as they say.

I rose early the following day, calling in at the newsagents to buy the *Daily Mirror* on my way to the lorry. The headline caught my eye: 'Now official, clothes off ration.' That was a laugh because the working class couldn't afford many clothes. Most people had to sell their coupons to rich people like Banksy to make ends meet. But still, it was nice to be able to shop without having to worry. There were two major events that happened in 1949: the nationalisation of road transport, and clothes coming off ration at last.

I checked the lorry and then jumped up inside the cab. Three hundred yards later I was pulling in at the main gate of Henley's. The people employed there were known to the Gravesenders as Henley's rubber-necks, because they made tyres, cables or anything made from rubber.

It took them quite a while to load me as there were so many different sizes of reels and cables which had to be first checked, then loaded. Three and a half hours later I pulled out of their yard.

My next stop was the Merry Chest for breakfast. I kept a low profile in the café. I didn't want to get too involved with other drivers and, this being my first long haul as a new driver, I lacked confidence. Also I had to work out my mileage and running times before I parked up that night. Finding transport digs was a worry too, so I didn't linger too long over breakfast.

When I left the café, I made my way to the A2 and headed towards the Dover Patrol, keeping a constant eye out for the wooden-tops because I didn't want to get nobbled and have my career put paid to once and for all. So I kept my speed to the legal limit of twenty miles an hour.

As I drove through London I kept a lookout for a signpost marked Brentford, on the A4. As soon as I saw the signpost I turned off, passing Gillette's on the way, and eventually ending up on the A30 for Basingstoke, which led onto the A303, the main arterial road to the West Country.

I engaged second gear most of the way as the road

was very hilly. By this time I was becoming tired. No-one realised just how tiring and stressful it was behind the wheel of an eight-wheeler – unless they'd done it, of course.

My actual speed after leaving Gravesend reached 28 mph, which was quite fast in those days. My eyes felt as if they were on stalks from the strain, and having to be on the alert all the time for wooden-tops didn't help.

My speed dropped considerably driving through the villages. I was beginning to nod and had to slap my face a few times to keep awake. I was really tired, my arms ached, and my right leg had started to go numb as I held the accelerator pedal down. Despite feeling tired, I was becoming more proficient at handling the lorry, and getting the knack of using the steering wheel correctly. My confidence behind the wheel was growing slowly but surely.

Driving through Mere, I started to work out my mileage and time in my head. I'd lost three hours loading and another half an hour booking on and off; that made three and a half hours. That left seven and a half hours' driving time. For my eleven hours, driving at 20 mph and allowing an average of sixteen, my time was due to be up within the next fifteen miles.

After a short while I came to a sign which read Wincanton twelve miles. I thought to myself 'thank goodness for that. My time's up when I reach there.' I had to devise a schedule of how to work out my time and mileage, even booking my digs in advance like the professional transport men. I'd never had to bother before because Fred always planned everything.

I stopped a couple of times to enquire where I could park the lorry for the night. I was finally given directions and fifteen minutes down the road I parked up alongside a heavy haulage lorry from Carpenter's Road, Bow. The driver was sitting in his cab doing his log sheets. What luck, I thought. I'll ask him if there are any transport digs in the area. I jumped down from the cab feeling quite saddle sore. My right leg still felt a bit stiff and the ball of my foot was hurting me. 'I don't know, I feel like an old man,' I said to myself.

As I walked around to the other lorry, the driver dropped his window. I asked him about digs. 'Be with you in a minute,' he shouted down. 'I'll just finish my sheets, then you can walk with me. I know of a good one not far from here.' I walked back to my lorry, made a log sheet out for that day, grabbed my night-out case and joined him. We chatted all the way to the digs.

The driver, Jim, was right about the accommodation: it was super. The rooms, especially the dining area, were more spacious than most that I'd stayed in previously. The landlady told me to call her Maud. The meal she prepared was really tasty, just like my old mum's. No sooner had I finished my dinner than other drivers began to arrive. I could tell by the way they joked with Maud that they were all regulars.

After listening to the wireless for a while, I joined three of the lads for a drink at the local pub. They turned out to be a great bunch. Jim wore a cheesecutter cap all the time; I hadn't seen him without it yet. Out of the three, he appeared at first to be the most strait-laced but turned out to be even more humorous than the other two. I noticed he didn't laugh at his own jokes, his face staying deadpan all the time.

The second driver worked for James Hemphill in Glasgow. He told us to call him Jock as everybody else did. What a comedian he turned out to be: a typical Glaswegian, he didn't give a damn. The third was Paul from Huddersfield. He drove an eight-wheeler AEC for Hanson's. They spoke about their experiences on the road, which I found captivating. But what surprised me was that they seemed to find me interesting too.

I told them a few stories about Fred and myself, and the antics I got up to as a trailer boy. I told them too that it was my first day on the road on my own, but my only concern at the moment was getting to know the distances and booking digs in advance. Paul told me to give it time and I would soon be on autopilot.

I told them I had to tip in Exeter. 'Well,' Jim replied, 'as you're going to Exeter, ring Frank Tucker. That's a large company. He'll more than likely give you a load back, then you might make Penton Corner café, Weyhill. It's this side of Andover. If you can't, you can always book another night back here.'

We got through another four pints and then headed back to the digs. As soon as my head hit the pillow I was asleep. My first day had been enjoyable but exhausting.

The following morning after a hearty breakfast, I waved goodbye to the lads. Jim called: 'Shay, if you do happen to get lucky with Tucker's and everything is straightforward with unloading and loading, your time should be up tonight round about Andover.'

It was a nightmare driving from Wincanton to Exeter because it was uphill most of the way, so I had to use third and fourth gear. It was virtually impossible to engage fifth. I could have done without all that hassle, plus it was pouring with rain which was literally bouncing off the road, making driving an arduous drag. Three long hours later I arrived in Exeter. I stopped at a nearby bus-stop and asked a feller standing there for directions to Alphington Road. It turned out to be past a railway station, just off Western Way – only ten minutes from where I was. At least something was going right for me.

The fellers there off loaded me within the hour, finishing at twelve noon. They very kindly looked up Tucker's telephone number for me as my fingers were like ice, it was so cold. The governor said I could use their telephone.

'What size lorry have you got?' asked the transport manager at Tucker's.

'Fifteen-ton eight-wheeler.'

I gave him the name and address of Banksy's company. Then he asked me to go to Heavitree and load out of West Brick, then deliver the bricks to Basingstoke. The manager said he would phone West Brick and let them know I would be there in about half an hour.

I jumped back in the cab, saying to myself: Heavitree here I come. The clouds were scudding across the sky in considerably brighter weather which made me feel happier. I didn't fancy loading bricks but it was my first return load and I felt really chuffed with myself.

West Brick was an easy place to find and in no time at all I was in the brickfields helping to load. Being an old hand at it I made sure that I was at the right end of the lorry because I remembered how the dust had got into my eyes the first time I loaded bricks with Fred. I'd now learned to be as crafty as them. When the work was done I roped and sheeted, then had a mug of tea and a sandwich before setting off again.

The time was now 2 pm. I had five hours left before my eleven hours would be up. I was now beginning to feel a part of the lorry, knowing exactly when to change down through the gears, and remembering to keep the revs up all the time because of the Gardner engine. On the flat road the other side of Honiton I managed to achieve 34 mph. I was moving effortlessly along, feeling as though I was on a rocking horse, dipping up and down. Having a long overhang over the rear wheels made the Atkinson rock. It was a great feeling.

After five hours which seemed like an eternity of hard slogging, I finally pulled in at Penton Corner, Weyhill . My hours were exactly eleven. Jim had been correct – he certainly knew his geography in the transport world. Penton Corner was a changeover café for night trunkers where the Londoners and Kent drivers swapped lorries with the West Country lads. They were a good breed who always managed to get through whatever the weather – snow, ice, fog or rain.

I was lucky to get a bed that night, as the place was packed with drivers who were completely worn out. But no matter how they felt about work and being away from their loved ones, the comradeship was incredible. I overheard one of the drivers talking about a driver nicknamed 'Donkey Dan' who drove an eight-wheeler AEC out of St Austell loaded with china clay and was known to be the best-hung man in the West Country. A python, they said.

We had had a good laugh that night and no offence was taken by anyone. It was just good fun. The West Country lads certainly liked a laugh.

I asked one of the staff in the digs to give me an early

morning call. At five-thirty sharp she rapped on the door of my room and called out, 'Wakey, wakey rise and shine'. After a good breakfast I was soon on the road. As I drove along, it suddenly occurred to me I hadn't derved up the previous night, so as soon as I came to a petrol station I pulled in and, using my agency card, filled up with derv. It was about twenty miles to Basingstoke, and it took me roughly one hour five minutes to get there.

The building site was just off the A30. It was one of the biggest that I had ever been on. It was so large it took me a quarter of an hour to find the section where I wanted to off-load the bricks. The foreman there sent the gang round to me. I asked one of them if he ever got bored off-loading bricks all day. His reply was that he did but they had a giggle sometimes, which made the time go more quickly. I didn't bother to ask what they were building. I supposed it was houses.

I had done really well that morning because an hour later I was driving out and making my way up to the A30, and once again heading towards London. I passed Heathrow and Earls Court, drove along the Embankment and over Battersea Bridge, passing the dogs' home along the way. I passed through Camberwell and New Cross, finally stopping at Blackheath, where I knew there was a wooden cabin selling fresh rolls.

After a 45-minute break I decided to ring Banksy. He sounded pleased to hear from me. He asked me for the telephone box number, saying he would ring me back. When he rang he asked me if everything was all right. I told him it was, that I'd managed to get a return load from Tucker's and had tipped at Basingstoke. That really pleased him because it was a good earner. He said to me, 'Shay, on your way back call in at APCM, Swanscombe and load some cement.'

'Okay, will do,' I answered, but privately I was thinking: Fuck it, that's all I need. Loading cement was really hard graft I could do without.

Forty minutes later I was wheeling her into the cement works. When I took a look at my delivery notes I was well chuffed, though, because it read Great Yarmouth. I would have to do a great deal of driving to get there, which meant less loading and off-loading. I didn't like doing local work with an eight-wheeler because that was sheer slavery, as you had to hand board it off. There was only one good thing about the cement works; everything was loaded by a conveyor belt, and the driver just sat in his cab. On the other hand this was another job which you had to hand board off. It was dirty, horrible work.

Soon I was driving down Banksy's lane. I drove straight onto the derv pump and started fuelling. I'd been standing there for about five minutes when Deafy came hobbling over to me, saying, 'Ssh' and putting his index finger over his mouth. He said, 'Don't say anything Shay, but there's been a lot of people here looking at the governor's property. Rumour has it he's selling up.'

I looked at him and whispered that I didn't think BRS would be interested in this place. He shook his head saying, 'No, no, no, not BRS. We all think it's private developers. But when you think of it, Shay, Banksy is getting on a bit, and let's face it, he's made his money. If he sells this ground he'll certainly make a few bob. Also, since nationalisation came into force the small businessman like Banksy doesn't know what the future holds.'

I agreed with Deafy, and said that in his shoes I'd probably do the same. 'Anyway,' continued Deafy, 'it doesn't affect me as I've only got three more months to go. But you, Shay, have got to get things into perspective and sort yourself out.'

'Don't you worry about me, Deafy. I'll certainly do that.'

'Good lad, good lad.' As Deafy walked away I heard him mutter, 'Things have never been the same since they got rid of steam lorries.'

I wandered over to the transport office. Kay looked very worried and subdued. I drew my own conclusions as to why. It must have been a shock for her too. 'Don't look so worried, Kay! It'll all come out in the wash. You'll see.'

She looked up, smiled and said, 'I suppose you're right, Shay.' I handed her my delivery notes from Tucker's and told her that I'd had two nights out but would hold on to the remaining two as I was off to Great Yarmouth in the morning.

She stood up and was about to tell the governor I was here when he popped his head round the door, saying, 'I thought it was you. Is everything all right?'

I told him I was loaded for Yarmouth and would be away soon, and he said I could take the lorry home if I liked, which I did. I blew Kay a kiss and left.

By now I had been driving on my own for two months. I had reached the grand old age of twenty-one and could now legally drive a heavy goods lorry.

It was about two weeks after my birthday when Sharpy told me it was definite that Banksy was selling out, and that poor Deafy was busy obliterating the names from the lorries that were already in the yard.

When I eventually saw the governor, he informed me that he had been offered a good price for the land and at the moment he did not know where road haulage was going to end up. Also, no contracts had been renewed, which did not help the situation either. But he told me he would give me an excellent reference when I needed it.

What grieved me most was the extra ten pounds he added on to my wages; I thought that was a bit paltry, considering the money he had made in the past. It wasn't the fact that he was selling up. Good luck to him.

'When do you want me to finish then, Gov?'

'Now,' he answered curtly, which I thought was unfounded.

I'm afraid my answer was: 'Well, all I can say is thanks for fuck all.' But as I walked away I was sick inside. That old lorry had become a part of me, and in a short space of time we'd covered some miles.

I moseyed on over to the garage to see Deafy who was standing there rubbing his hands with glee. I heard him tell the others, 'I'll be finishing here two weeks earlier than I expected.'

The other drivers didn't seem bothered either, as they had seen it coming. Dave remarked, 'Well, lads, that's definitely made up my mind. I'll be off to the valleys. That'll please the wife as her folks are all in Wales.'

We all shook hands. Then Sharpy put an arm round Deafy's shoulder, saying, 'All the best to you.' I saw that Deafy was a little choked at this gesture. It surprised me really, because those two were always bickering with one another. At times the names they called each other were unbelievable.

I grabbed my night-out case from the cab, shook hands with Kay and pecked her on the cheek. She promised she'd post on to me all that I was owed.

I asked Deafy to give my best to all the other drivers when he saw them. He had to stay behind because the governor had asked him to finish off the lorries, as apparently a company from Yorkshire had bought them and was sending a few drivers down to take them back. If the governor had asked me to finish painting the lorries I would have told him to get lost.

Sharpy, Dave and I walked up the lane. Suddenly Sharpy piped up, 'I thought the old man would have given us a lift in his car, tight old bastard.'

Although I was upset, Banksy had done me proud really, as I now had a licence. Fred had taught me what he knew about road transport, so I had become an experienced driver.

All the same, it was the end of an era.

Chapter 9

Arnold Transport

WHILE I was waiting for the 480 bus to Gravesend, one of Jock Mullock's workers came along, driving a long-wheel-based, snub-nosed Morris Commercial. I immediately held my hand out. Seeing the way I was dressed, he pulled up. I ran up to the nearside and jumped up inside the cab.

'You're Phil Purdon,' I said to him. 'I've seen you many times in the Merry Chest. Your dad works for Arnold's like mine, doesn't he?'

'Yes,' Phil replied. 'That's right.'

I told him that I had just been made redundant from Banks and would be looking for work.

'I don't know if you're interested but Jock wants a driver. It's all market work, though, a bit different from what you're used to. It would tide you over until something better comes along.' I was definitely interested, and he promised to ask the governor for me.

I went with Phil to the yard. Jock Mullock was a stockily built man of medium height and with jet-black hair. Phil shouted over to him, 'Do you want any drivers, Governor?'

He did.

'Well, I've got someone here who's been made redundant. He drove the heavy stuff for Banksy's Transport.'

'Oh, I'd heard on the grapevine that he was packing it all in. It's true then?' he asked, looking at me. He had a very broad Scottish accent but spoke clearly.

I confirmed that Banksy had sold up.

'So, you're looking for work? Well, be round here at 2 pm sharp tomorrow. The other drivers will be here then, so they'll show you the ropes.'

I thanked him and as I walked out of his yard I thought to myself: blimey that was quick. I've only been out of work an hour and I'm back in tomorrow. Can't be bad.

All Jock's lorries had 'B' licences, so the drivers could only do a 25-mile radius from the depot. His yard was opposite Gravesend railway station, so getting to work was no problem for me. It'd certainly tide me over for the time being.

It didn't take me long to learn the ropes on Mullock's. This job was all night work. All the drivers had mates. The young lad who came out with me was sixteen years old. Having been a mate myself a few years back, I knew the feeling only too well. It was nerve-racking, so I took him under my wing, as it were.

One o'clock in the morning loading lorries with cabbages by moonlight, in the middle of a field, was hard work. They had to be delivered to Covent Garden in London early in the morning.

What I liked was that everybody mucked in, which made life a lot easier. There were ten of us all together: five drivers and five mates. Phil Purdon's mate, Vic Parrot, was a powerful seventeen-year-old, strong as an ox. We had loaded a lorry with bags of greens one day, when he threw a bag up in the air and it landed the other side of the lorry without even touching the other bags. Another time when we were unloading in the market, he put one ton of potatoes on one of the porter's barrows, pulled it along the road to the stand with no effort at all, then unloaded it. Fifty years on, Phil Purdon still talks about it.

We were now in the late 1950s, and after a while I felt as though I was stuck in a rut. Time was passing only too quickly. Phil Purdon left and got a job working for Ridgewell's at Northfleet,driving an eight-wheel AEC lorry delivering lead to all parts of England. Phil doing that really made me get my arse into gear. I scanned the papers daily. I saw a number of vacancies at Arnold's, driving eight-wheel Atkinsons up and down the A2.

The gentleman – Mr Chase – who interviewed me was very quietly spoken. The drivers always called him by his surname. I had been told by different people that he was a religious man who did not like drivers blaspheming in front of him. His wishes were always respected. Yet his sense of humour was terrific. I suppose it had to be, since he was dealing with transport men.

After the interview Mr Chase told me I could start the following week. I was delighted to be given the opportunity of being back driving my favourite lorry. Naturally I would be driving one of the old ones, Fleet No. 29. In fact I remembered Kenny Campbell driving it years previously. He was now driving the latest model.

Mr Chase was the transport manager, but Joe Atkins was the governor of the company. It was one of the largest companies in the Medway towns. Mr Joe, as he was known, left the running of the company solely to Mr Chase.

Arnold's Transport was an old, well-established company, situated at the Lion Garage, Gravesend. The lorries were always fully loaded each day. British Road Services were not interested in it because it was a spot hire transport company, so it didn't have any contracts. Although Arnold's did have some 'A' licences, it didn't make the slightest difference. But in the days of BRS, companies like Arnold's didn't know what the future held. That's why Arnold had put his company up for sale.

It was the Atkins brothers who bought Arnold's transport out. Joe Atkins always had ERFs before they nationalised him. Arnold's had Atkinson's lorries, and as they were up to Joe's standard he continued buying them. In time he had the largest fleet of eight-wheel Atkinsons in Kent. Mr Chase operated the company one hundred per cent legal. If a driver's time was up it would mean a night out for him. Even if he was only a quarter of an hour away from the depot, a night loader would collect it, load it, then drive it back to where it was originally. It took some drivers by surprise in the morning.

The Atkins brothers' fleet was their pride and joy, and every four years the lorries were stripped down to their chassis and rebuilt.

We heard Joe Atkins had got his Gyproc contract back from the BRS, so new lorries were arriving regularly. He was only permitted 'C' licences on them so fought a long and bitter court case from which he walked away with over fifty 'A' licences. It was a topic of conversation in the road industry for months to come.

My first trip was to Plymouth. I found out later that there would be four others going to the same place, me being the fifth. The lorries had already been loaded for us by the night loaders.

We pulled out of Gyproc at six in the morning. The first was Ron Percy, followed by Kenny Campbell, Jock Sterling, Ron Cannon and me. Although the other four had later models, the engines were all Gardner LWs. They were capable of thirty to thirty-four miles per hour. The actual speed limit was still 20 mph.

The four of them drove swiftly through London, weaving in and out of the traffic, handling their lorries as if they were saloon cars. They made those lorries sing driving through the streets.

Our first stop was Jock's café at Colnbrook, near Slough. While we were sitting there eating our breakfast, one of the drivers told me that our time would be up when we reached Ilminster and that's where we would book off on our log sheets. However, we would drive on until we reached Honiton then we would have time for a nice lie in the following morning.

'Your lorry goes well, doesn't it, Jock?' I said to him. He fell silent, then he and the others gave a slight grin. I had guessed right. They were all driving with their bar pulled back, which was illegal really.

As it was four hours' drive to Wincanton, on the A303, one of them suggested we stop at the Frying Pan for a break and also book our digs. As soon as I jumped up inside the cab, I lifted up the bonnet and tied the bar back, so that I would over-ride the engine. However, as I had learned as a trailer boy, this made the accelerator harder to push down.

I flicked the pilot switch down and started her up. As

I looked out of the windscreen I couldn't see the other four for dust. The race was on, but inside my head I knew I had to be extremely careful, because if the brake drums got hot, the vacuum brakes wouldn't hold me back and it only had six-wheel braking power.

The A30 was very narrow. Arnold's transport was such a well-known company that when the West Country drivers spotted us in the distance, they flashed their lights.

As soon as I started to go down the hill, I knocked the gear lever into the silent six. I used all my skill and ingenuity to try and catch them up. I even took chances that I would not have done normally. As I passed the Tower café, Basingstoke, I caught sight of the other four in the distance. It had taken me two and a half hours to catch them. We played cat and mouse all the way to the Frying Pan café. As I was about to turn off the A30 and onto the A303, I saw Phil Purdon and his father, both driving AEC lorries, on their way back from Southampton. I waved as I turned off onto the A303. I presumed they had just delivered their load, which was lead, to Pirelli's Cables. Phil Purdon Senior, along with many other drivers, had worked for the original Arnold as well as Joe Atkins. But it didn't matter which company a driver worked for; they all knew one another.

As we passed Weyhill, two of Arnold's drivers were coming in the opposite direction, one being Fred Coe, the other Joe Waterman. At that precise moment I counted seven Atkinson lorries belonging to Arnold in the same area. A thought suddenly occurred to me: some companies would have given their eye teeth to own two, let alone a fleet.

The driver who was leading started to slow down to let the brake drums cool. If you failed to do this the brakes would completely fade out, and if this happened the lorry would actually run away with you, which was very dangerous. A driver gained knowledge and experience over the years. Knowing your vehicle was very important.

I was beginning to feel jaded as I pulled into the Frying Pan café. As I climbed down from my cab, Ron Percy greeted me with a grin on his face. He looked how I felt. He was stockily built with a round, cheerful-looking face and a thick handlebar moustache which gave him the appearance of an RAF officer – which he wasn't.

His sidekick, Jock Sterling, was a genuine lad, a very likeable person who got on with everyone. The only trouble was he had a very short fuse.

Kenny Campbell, from a large family of transport drivers, was a favourite in the driving world because you always knew where you were with him. He didn't mince his words and always spoke the truth. Most of the time he was on the same wavelength as me, two individualists.

Ron Cannon from Gravesend was a dependable comrade to his fellow workers. He always gave a hand when needed, such as with roping and sheeting, but if he thought something or someone was out of order he would have his say and wouldn't always back down. If you were in an argument and were right, he would back you all the way.

We had a light lunch as we were all having an evening meal later on that night. The journey from Rochester to the Frying Pan café had only taken us seven hours which was excellent. Ron Percy walked to the telephone booth in the corner of the café and booked us all in at the Cosy café which was in the main high street of Honiton. Jock said we had better make a move as it was still a long drive to Honiton. We had all agreed that the Cosy café would be the place to stay for the night. Kenny Campbell took the lead, driving like a bat out of hell. Some of the villages were so small that if there had been an emergency it would have been impossible to stop in time, especially having vacuum brakes and being in the silent six.

I could remember doing this journey only too well when I was working for Banksy. The lorry I drove then had an 'A' licence, but now I had one which carried a 'C' licence – no return loads home. At one stage all I seemed to be doing was acknowledging the West Country drivers on my way through. Arnold's were so well known in this part of the country.

After three and a half hours we arrived in Honiton. Altogether we had been driving for ten and a half hours, covering a distance of about two hundred miles.

The six of us ambled across the road from the Cosy café to the pub.

A sixth driver was there: Don Hillier from Wales, who was driving home empty. He had worked for Arnold since the Second World War ended, during which time he had met a young lady from Gravesend and decided to settle in Kent.

We had a game of darts with the locals. Kenny Campbell took charge of the scoring and that meant only one thing: we couldn't lose. He was a crook if ever I saw one. I had to laugh at him. The funny thing was, nobody sussed.

With all the different accents it sounded like the League of Nations, especially when Jock got excited – no-one could understand him at all. My opinion, for what it was worth, was that the Gravesend accent sounded more cockney than the East Enders themselves.

Although our running time was early, we agreed to leave punctually the following morning. Also, we had decided to spend our third night in our own homes. Of course, if a driver was caught it meant instant dismissal.

As we left Exeter, it was crawler gear all the way, because we had to go up Telegraph hill which was very well known to lorry drivers. It was a long drag all the way up, working the Gardner engine hard.

Three and a half hours later we finally arrived in Plymouth. Everywhere you looked there were new buildings being built; the site where we parked was enormous. The workers there placed plaster boards along side of the road to make it easier for them to off-load. It only took them three hours, because they had their own fork-lift, which really surprised me.

The five of us agreed that we would stop off once more for a break, this time on top of Telegraph hill where there was a transport café. While having tea, Ron Percy said he would go and phone the Cosy café, Honiton and book us in again.

He told us our time would be up at the Jubilee café in Bexley tomorrow, 'unofficially, of course,' he said with an artful smile. 'So we'll book off then, which will put us on a dodgy night.'

Ron piped up, 'That's settled then. I'm taking my lorry home. Don't know about you lot.'

'It won't matter what we do, or where we park, we'll still book off at Bexley,' Percy replied. Then Kenny told us that Tom Atkins had asked how he was getting on with his new lorry. He told him it was all right and that he liked it, but had to see Mr Chase about his insurance because he'd been caught speeding. The governor asked what speed he was doing. Kenny said, 'And I answered, coughing, "Only twenty-five miles an hour." '

Jock said, 'Never mind. The dodgy night will help pay for the fine.'

The next day the five of us drove hell-for-leather to get finished, stopping only once at the Towers, Basingstoke. When we pulled in, other Arnold lorries were parked there already. But then of course we got carried away chatting, lost time and had to make it up again.

Kenny Campbell's lorry had no sign-writing on the cab. The sign board on the top was a reversible one, having Arnold on one side and Gyproc the other. So the company wasn't sure whether to put an 'A' or 'C' on the vehicle. It became a contract 'C' lorry in the end.

My first trip working for Arnold went very well. I was really beginning to feel part of the company.

I was now driving up and down the West Country regularly. Kenny Campbell left the company and went to work for Harold Woods Tankers. Later, Ron Percy and Jock Sterling followed suit.

I remember so well a driver named Bill Wheeler, a very experienced eight-wheeler driver. He'd joined Arnold in 1958 and was very well known in the transport world. Like most of us he had worked at different companies. Within a year of joining he became shop steward, and would have been the best in the country if he hadn't been so militant with everybody. For reasons only known to himself he was very anti-management.

The first month he was shop steward, he complained so much to the governor about the Atkinson lorries being cold that in the end Arnold's issued all their drivers with tailor-made overcoats with white bands on the sleeves. This had the advantage that the drivers' hand signals were easily seen in the dark. In no uncertain terms Bill told the management that the drivers were not children, they were adults, and had to be treated as such.

Finally the running times were altered too. Although Joe Atkins was a very clever businessman, Bill Wheeler had him worried because he didn't know what was coming next. Bill's downfall came through not knowing when to stop. When he joined he was given a brand-new fibreglass Atkinson, but after a while he was not satisfied with that, so asked if he could work in the yard shunting lorries, which was granted.

If a driver made a mistake, Bill always covered up for him, then turned the tables so it looked as if the management was to blame, making them look incompetent. But the management were fully aware that he had a lot of backing from the drivers, and were fearful of losing their contracts because they knew for a fact that Bill Wheeler would have them all out on strike if the need arose.

One day Bill called a meeting at Gyproc, and because the toilets were centrally heated, it was held in there. Mr Chase heard of this meeting and joined them, but he was promptly told to 'clear off'. Two weeks later another meeting was held, this time at the back of some lorries. Afterwards Bill went to the management and said that, as the national speed limit had gone up to 30 mph, the drivers would only keep to it if they were given a fifteen per cent pay rise. They would work to an average speed of 20 mph – which the management accepted.

It was then that Ben Tillet decided to have a vote on whether to keep Bill Wheeler as shop steward or not. It was a unanimous decision, with all the drivers deciding on the latter. The management, I must say, was over the moon.

The very next day Bill was given five deliveries to London with an old eight-wheeler that was bloody hard work. Talk about adding insult to injury. We watched secretly as Bill jumped up inside the cab and moved off. We could see he was fuming.

We heard on the grapevine that he drove to the Woolwich Ferry, telephoned the Ministry of Transport and asked them to send someone out to him, which they did. He told them that the lorry was unfit for the road. Then Bill left the poor inspector standing there while he marched off in a temper. Apparently he hitchhiked home and has never been seen again to this day.

The company went on to expand more and more. One great step forward was when Joe Atkins ordered the latest type 150 Gardners which had full eight-wheel air brakes. The steering post had been shortened and they all had crash bars on the front. They came home nine at a time, the last batch of eight-wheeler lorries to arrive at Lion Garage, Gravesend.

As the old drivers left, more new faces appeared. One was a very quiet young man called Alan Spillett who later became one of Britain's leading trucking artists. He still paints beautiful road scenes and lorries in all parts of the country, and his work is in great demand.

One Sunday six drivers left for the West Country and the following Monday another twenty-five followed on. Eleven of them went to Bristol. Word went out, unofficially of course, to try and make Bristol.

On Tuesday night I counted thirteen Arnold lorries all parked in Midland Road, Bristol. The car-park attendant, a lofty-looking man who knew most of the Arnold drivers, couldn't believe his eyes.

This man was very athletic and fanatical about football. We gave the young local boys from his club a game one day on the National Car Park. They were dressed in their usual red shirts and white shorts. To us they looked like a bunch of pansies. They soon proved otherwise. However, although they were quick on their feet, a few crafty nudges from us soon slowed them down and the game ended in a draw. They were smashing lads who took all our cheating in their stride. As for the car park attendant, he thought it was the best game he'd seen in ages. That night nobody paid for parking.

Just a short distance from where we were staying there was a night-club called the Carlton, which was open to the small hours. We smartened ourselves up, and spent a good many hours in there. The beer was flowing fast. The tables were laid out in a circle, so the customers had a great view of the girl who did a strip-tease. It was tastefully done and she was a gorgeous girl but we found it so boring. Most of us got up, walked over to the bar and had a few more pints. After a while the manager came over and asked us to leave, as in his eyes we were becoming too raucous – which I suppose we were.

It was a good thing the manager turned us out when he did because when we came back to Gravesend, we found out that Mr Chase had had to bail six of his other drivers out of the nick in Exeter to get his lorries home. In fun on the quayside road works they had turned most of the signposts in the wrong direction, causing car drivers to go up dead ends. It was mayhem.

Time moved on. It was now 1968 and I was driving the latest eight-wheel Atkinson. It was altogether different from the others I had driven. The cab was made entirely out of fibreglass. It had a six-speed gear box and the latest 150 Gardner engine. All eight wheels were fitted with a full air-braking system. It was a dream: 12 mph faster than the LW used to be.

I left the depot at Rochester one day for Southampton. As I drove up towards Swanscombe cutting, I suddenly had the feeling that I was being pushed from behind, but realised that the new engine was pulling much better than the ones I had been used to. The LW engine felt really antiquated.

As I drove along the A30 the speedometer was just reaching 47 mph. What a superb lorry, I thought to myself. The Atkinson's steering post was now lower, which made it steer a lot better when negotiating corners. Having a sixth gear made a great deal of difference too. It was known as the 'overdrive'.

Feeling peckish, I pulled in at the Towers in Basingstoke for breakfast. Inside were a few Arnold's drivers. As I sat down to join them, Bob Shersby and Noddy Durling were laughing. Keith Austin, who was a great character and didn't care a monkey's uncle about anything, was telling jokes. He was reeling them off. I don't know how he remembered them all.

As we walked out onto the lorry park the jokes were still being bandied about. Although the drivers were nearly always away from home, roping and sheeting in all weathers, and short of money, they still had a wonderful sense of humour and could always laugh at themselves.

After leaving the Towers, I arrived in Southampton. Within the hour, I was unloaded by a fork-lift driver in the builders' merchants. I managed to get a return load of apples to be delivered to cold storage at Marden, in Kent. I had to load out of Pitters Transport. I had just finished roping and sheeting my load, when one of Arnold's night trunk lorry drivers pulled in from Southampton. He wound his window down and shouted down to me, 'Where are you stopping tonight, driver?'

'Southampton,' I replied.

'I'll take your lorry back tonight on trunk, if you like,' he said.

Looking straight at him and with a gleam in my eye, I replied in a quiet voice, 'I've just loaded fifteen tons of apples, roped and sheeted. Now you want me to do the same to your lorry, then drive your old LW home tomorrow? Come off it! Good try, though.'

I could see he was pissed off. 'Well, if that's your attitude . . .' he replied. I said no more but pressed the starter and left the yard. I pulled up outside a boarding house called Rose's which was used by all Arnold's drivers. The woman there always served every driver a large thick steak, whatever time of day it was. I often wondered how she made it pay. She was a lovely lady, always had a smile. The drivers teased her unmercifully. But she loved it and took it all in her stride. As far as I know, she lived on her own.

Jock Watson and Bob Shersby, drivers I knew, were in there that night. We got talking and decided to go to the 'dolly mixtures'. It made a change from standing at the bar.

Bob's lorry looked swish, painted in the new Arnold colours – black and white. When I mentioned the night trunker to them, they laughed and remarked, 'You have to watch those bastards. If you give them an inch they'll take a yard. He's only got to drive to the Staines by-pass, then he changes over with a Kent driver anyway.'

The following day we got up early, arriving in Marden around noon. The three of us soon got cracking, throwing the boxes of apples down a conveyor belt. The time just sped by. In total we had shifted 45 tons of apples between us, which was bloody hard work.

Jock Watson rang Arnold to see about more work. He was told that Bob and I had to load out at Reed's of Maidstone, then deliver paper to Manchester. Jock had to load out of Imperial Paper Mills in Gravesend, then onto Shand Kydd's at Christchurch to off-load.

When Bob and I arrived at Reed's Jock Gillespie, Brian Burns and Keith Austin were already there loading for the same place. Jock was now driving one of the new 32-ton articulators.

It was all crane loading at Reed's, which was far better than many other sites. Unfortunately, this particular day was very windy, the gusts of wind creeping under the sheets and blowing them off. So we made the decision to rope and sheet one lorry at a time. There was nothing worse than trying to do this on a blustery day.

As we left Reed's we drove in convoy to Cory's Road, Rochester, where we parked for the night. There must have been at least forty Atkinson vehicles. It certainly was a large transport company. Whichever way you looked, there were Arnold's lorries.

We reported to the night shunter, Vic, from whom we had to collect our night-out money which by now had gone up from seventeen shillings and sixpence to one pound. My mind flashed back to when I used to stop at Mrs Deacon's boarding house which was near a railway station in Derby. She had trained her parrot to greet drivers with 'Bed and Breakfast, ten shillings and sixpence.' It created a laugh but when she put her prices up, the bird still shouted out the same, which confused a lot of drivers.

That day when we all trooped in for our money, Vic told us that he'd spent the float, which happened occasionally and wasn't at all unusual for him. On one particular occasion Bob called him a prat, and asked, 'How are we supposed to survive without any money? We'll have to ring Mr Chase and report you.'

Bob turned round to us and winked. Vic's face was a picture; he really freaked out, saying: 'Don't do that! Don't do that!' He dug down deep into his pockets and pulled out some notes, complaining, 'But I won't have enough for the other drivers in the morning.'

'Why do you think we're collecting our money tonight?' Jock Gillespie replied.

Vic was a hard working man, though. Most nights he shunted, loaded and sheeted an average of forty lorries.

The following morning we all arranged to meet at five-thirty sharp. The five of us drove out of Rochester and headed straight for the Merry Chest café where June cooked us a hearty breakfast as usual. She certainly knew the way to a man's stomach. She knew most of Arnold's drivers by name. How she remembered them all was beyond me. Her husband Ben was a great character too. They were like family to all of us.

The café was later handed down to June's children, but the old familiar welcome is still there. The Merry Chest is one of the last remaining original transport cafés in the country.

We bade our farewells to June and thanked her. One by one the five of us pulled out of the car park. We drove through London onto the M1 but stopped again for another break when we reached the Blue Boar.

Jock Gillespie and Brian Burns were noted for playing the fool, but it was nearly always Jock who had the devil in him. He seemed to get a great kick out of playing cat and mouse on the trunk roads.

As we were delivering just outside Manchester, Keith mentioned some digs that he knew near Stalybridge. So we stayed in convoy behind him and followed onto a disused piece of land to park up for the night.

After bathing and putting our best bibs and tuckers on, we decided to see what the local Chinese restaurant was like. As we sat at the table, some local children started peering in and banging their fists on the window for attention. They chanted at the top of their voices: 'Chinky, chinky Chinaman', then ran off giggling. The owner, instead of ignoring them, chased them, which made things worse. They thought he was playing at first, but the second time they appeared, the kids got the shock of their lives because he lay in wait with a bucket full of cold water. They didn't think that was so funny and bolted. They didn't return, not while we were there anyway.

We left there and sauntered across the road to the local pub. Poor old Bob had broken out into a sweat. 'What's the matter with you?' I asked, turning round.

'I can't afford to buy any drinks,' he replied.

'Don't you worry about that, you silly old sod. We know you're strapped for cash. We're not exactly flush, but I think we can run to a pint,' I replied. I felt sorry for Bob because it was his commitments at home that made him short of funds. Jock pushed open the door and shoved him inside, saying, 'Cheer up you miserable old sod.'

After two or three pints he soon forgot his troubles, and once we got him laughing he was good company. Brian Burns didn't give a toss about anything. He was a young, lean-looking man, a good character to have around if you felt a bit down.

Arriving back at the digs, and having had quite a bit to drink, we undressed for bed. Jock, who I might add was the first to strip off, was just about to get into bed when Burnsy, looking straight at him, shouted, 'Ah, crumpet!' and jumped up. Jock, seeing what was about to happen, moved aside, and Burnsy, who was six foot, hit the floor with a horrible thud. We thought he had broken every bone in his body but no, he just lay there laughing. I think he was so drunk that he must have bounced.

A pillow-fight was next on the agenda, feathers flying everywhere. Burnsy was jumping up and down on his bed like a two-year-old. It's a wonder it didn't break in half.

Keith, being just as bad as the others, picked up a tin wastepaper basket and threw it down the stairs. We heard it clanging all the way down to the bottom. It had a long way to go as our room was three flights up. Suddenly, when it had quietened a little, we heard the landlady's door slam. Immediately we all jumped into bed like naughty little schoolboys, pretending to be asleep.

As we lay there, the feathers were still floating down from the ceiling. I smiled to myself and fell asleep.

As we ate our breakfast the following morning, we expected the worst but when the landlady returned to collect our dirty plates, she didn't mention anything to us. Obviously she had not been up to the room. No sooner had we finished than we made a quick exit, not wanting to stay around to face the music. We felt really guilty as it was and sorry for what we had done. Nothing was broken but it looked just as if a bomb had exploded in the room.

Directly we got out of there we almost ran towards our lorries. As we stood talking for a while, Gillespie said, 'Do you remember the time when four of Arnold's drivers stopped in Midland Road, Bristol?' None of us did.

'Oh, you will when I start telling you. The building, if I remember rightly, was very old.' As Jock told the story, it started coming back to us.

One night, as Bill Draper was preparing for bed, we had started to tease him unmercifully about his underpants. They looked like Joseph's coat of many colours. One of the lads wolf whistled and shouted: 'Wow, look at those legs!' Then he rushed across the room, and jumped on top of Bill, who fell backwards onto the bed. He landed on it so hard it collapsed, and one of the legs of the bed went straight through the floor and the ceiling below. Neither of them could move for laughing.

The landlord, on hearing the crash, came running up the stairs, yelling, 'What's going on up here?' As he entered the room he just glared at the two of them, one lying on top of the other. You could see the disbelief and horror on his face.

Bill got up calmly and, looking him straight in the eye, said, 'Sorry. It's not what you're thinking!' How he said it with such a sober look on his face, I'll never know.

'Well, to be quite honest I don't know what to think,' the landlord replied. Anyway, we pacified him by saying we'd give him some plasterboard and help him to repair the ceiling. It was the least we could do.

The following morning they were true to their word and got the job done. It looked better than before because they painted the ceiling as well. There were a few sheets left over, but goodness only knows what the landlord did with them.

Years later, Bill Draper jumped up onto the back of his unit and stepped on the battery box, hurting his leg. He continued to work for a while, but his leg got progressively worse and eventually he had to have it amputated. He later died, which was very sad; he was greatly missed.

The paper mills where we had to unload our reels were about a quarter of a mile away, so it didn't take us long to get there and tip. We pulled alongside the road where Keith had noticed a telephone box. Jock rang all the places he knew in the Manchester area, even

Sutton's of St. Helens, to get a return load home but was unsuccessful. There was plenty of work to be delivered further north but none of us wanted that. Jock definitely had the devil in him that day because he opened the telephone box door and shouted to the others: 'There's no work for the south but there's plenty going to Scotland. Any offers?' It all went silent. So Keith decided to book us all in to load out of Manchester to Scotland.

Bob Shersby went ballistic. 'Scotland!' he shouted, pulling the receiver from Keith's hand. I watched as Jock Gillespie actually slid down the side of the telephone box onto the ground, clutching his stomach: he was in pain with laughter.

As I sat in my Volvo 88 parked up at Markyate, I smiled to myself as I recalled episodes in my past. Those were the good years for me as a transport driver. Now, in 1975, it just felt as though all that had come to an end.

Drivers are a new breed today. There are lorry drivers and truckers but, in the good old days we were classed as transport men. In my mind, I selected the real transport drivers from the rest, as they ambled towards the café. There are still a lot of them left out there. To me they are all Fred Ruddocks.

Other Titles from Old Pond Publishing

Juggernaut Drivers

It's the 1970s and 80s. In Les Purdon's entertaining novel a group of owner operators do their best to make a living out of trucking. Sometimes they run legal. Sometimes they don't.

Alaska Highway TRUCKSTAR

DVD showing owner-operators driving under pressure for timber and oil in extreme conditions from 50 Celsius below to mud and soft going.

Heavy Transport TRUCKSTAR

Seven exceptional loads, mostly of 100–460 tonnes but including a salvaged U-boat filmed in Europe and the United States.

Custom Cutters DYLAN WINTER

Two combine crews filmed through the harvesting season as they head north from Texas towards the Canadian border.

Farmer's Boy MICHAEL HAWKER

A detailed account of farm work in N. Devon in the 1940s and 1950s. Paperback

Early to Rise HUGH BARRETT

A classic of rural literature, this is a truthful account of a young man working as a farm pupil in Suffolk in the 1930s. Paperback

A Good Living HUGH BARRETT

Following on from *Early to Rise*, Hugh takes us back to the assortment of farms with which he was involved from 1937 to 1949. Paperback

A Land Girl's War JOAN SNELLING

Work as a tractor driver on a Norfolk fruit farm and wartime romance vividly recalled. Paperback

Land Girls at the Old Rectory IRENE GRIMWOOD

Light-hearted, boisterous memories of land girls in Suffolk 1942–46. Paperback.

Free complete catalogue:

Old Pond Publishing, Dencora Business Centre,
36 White House Road, Ipswich IP1 5LT, United Kingdom
www.oldpond.com Phone: 01473 238200 Fax: 01473 238201

About the author

Leslie Purdon was born in Gravesend. His only ambition from infant school onwards was to become a transport driver like other members of his family. During the last months of his school career he played truant in order to haul hardcore in a five-ton Canadian Ford on private land from Gravesend aerodrome to a local building site. After leaving school he became a driver's mate, got his driving licence and was called up for national service.

When Les was demobbed he drove eight-wheel Atkinsons on an 'A' licence, working for Arnold's Transport at Gravesend. He went on to drive for Everard's road tankers, Tankfreight, Pickfords, Thomas Allen and for fourteen years with BP long distance. His final stint was with P & O before he retired in 1998 after 46 years in road transport.

After he retired, Les recalled his early working years in *Truckers North Truckers South*. He followed this with *Juggernaut Drivers*, the story of owner-operator Dennis Richardson and his chums trucking in the 1970s and 80s.

Much of his spare time now is taken up with his 1946 OLAD Bedford which he and his wife Pauline take to many steam and road truck shows. He lives in Lordswood, Chatham, Kent, where he is known as 'the man with the vintage lorry'.